A TIME TRAVEL DIALOGUE

A Time Travel Dialogue

John W. Carroll, et al.

http://www.openbookpublishers.com

© 2014 John W. Carroll

This work is licensed under a Creative Commons Attribution 4.0 International license (CC BY 4.0). This license allows you to share, copy, distribute and transmit the work; to adapt the work and to make commercial use of the work providing attribution is made to the authors (but not in any way that suggests that they endorse you or your use of the work). Attribution should include the following information:

Carroll, John W., et al., *A Time Travel Dialogue*. Cambridge, UK: Open Book Publishers, 2014. http://dx.doi.org/10.11647/OBP.0043

In order to access detailed and updated information on the license, please visit: www.openbookpublishers.com/isbn/9781783740376#copyright

Further details about CC BY licenses are available at:
http://creativecommons.org/licenses/by/4.0

Digital material and resources associated with this volume are available at: www.openbookpublishers.com/isbn/9781783740376#resources

ISBN Paperback: 978-1-78374-037-6
ISBN Hardback: 978-1-78374-038-3
ISBN Digital (PDF): 978-1-78374-039-0
ISBN Digital ebook (epub): 978-1-78374-040-6
ISBN Digital ebook (mobi): 978-1-78374-041-3
DOI: 10.11647/OBP.0043

Cover image: *Bubble Chamber: Antiproton Annihilation* © 1971 CERN, all rights reserved.

All paper used by Open Book Publishers is SFI (Sustainable Forestry Initiative), and PEFC (Programme for the Endorsement of Forest Certification Schemes) Certified.

Printed in the United Kingdom and United States by Lightning Source for Open Book Publishers.

Contents

List of Illustrations	vii
Introduction	1
1. Monday	3
2. Tuesday	15
3. Wednesday	33
4. Thursday	45
5. Friday	57
Notes	77
Credits and Acknowledgements	79

List of Illustrations

This book contains animated graphs. If your device supports MP4 video files, please click on the image to trigger the animation. If you are reading a printed edition, or on a device that does not support videos, please scan the QR code in the relevant footnote or visit www.openbookpublishers.com/isbn/9781783740376#resources to watch the animations.

1.1	The Unanticipated Results	5
1.2	The Anticipated Results	7
1.3	Backwards Causation?	9
2.1	A Time-Traveling Psi-Lepton?	15
2.2	Willie Slips Up	23
4.1	A Causal Loop?	47
4.2	The Really Weird Results	54
5.1	The Life of Tad	60
5.2	An Ordinary Psi-Lepton: No Time Travel	64
5.3	A Time-Traveling Psi-Lepton as seen from the Departure Branch?	65
5.4	The Disregarded Results	66
5.5	A Time-Traveling Psi-Lepton as seen from the Arrival Branch?	68
5.6	Trigger-On Departure, Trigger-Off Arrival, as seen from the Arrival Branch?	70
5.7	A Challenge for Tad's Multi-Dimensional Hypothesis	73

Introduction

> Any question of philosophy, [...] which is so *obscure* and *uncertain*, that human reason can reach no fixed determination with regard to it—if it should be treated at all—seems to lead us naturally into the style of dialogue and conversation.
>
> —Pamphilus to Hermippus, from David Hume's, *Dialogues Concerning Natural Religion*

Theoretical physicists take seriously the idea of time travel; some, including J. Richard Gott and Paul Davies, have published monographs, accessible to the layperson, describing the extraordinary work that has been done. Experimental physicists have on occasion even come face to face with the possibility that time travel to the past is real. This book is built on conversations set in 2010 at the Jefferson National Laboratory where unanticipated data led distinguished physicist Dr. Carlene Rufus to investigate a hypothesis of backwards time travel.

The conversations are of philosophical interest. In addition to Dr. Rufus, the other participants in the conversations are Tad Logan, a graduate student research assistant, and William Esquire, a philosophically inclined computer scientist. Their careful, and often humorous, thoughts wander from the experimental data, to science fiction cinema, and even to thoroughly abstract and metaphysical paradoxes about the permanence of the past, the privilege of the present, the nature of causation, and what one can or cannot do. The question of the possibility of time travel is a truly abstract and multifaceted one. So, it should be no surprise that philosophy, with its diverse subject matter, speculative powers, and its reverence for logic, so grounded as it is in the tension between the familiar and the unfamiliar, should here rear its head. Indeed, it is Willie's philosophical input that shapes both the planning and analysis of Dr. Rufus's experiments.

At the beginning of Hume's *Dialogues Concerning Natural Religion*, Pamphilus reports to Hermippus of having had the occasion to observe the insightful conversations of Cleanthes, Demea, and Philo. It is with similar good fortune that the conversations of Dr. Rufus, Tad, and Willie can now be made public. Thanks to detailed notes and well organized data, one week of our researchers' scientific work and philosophical discussions is readily presentable, with tolerable accuracy, in the dialogue format long upheld by Plato, Berkeley, Hume and so many others. Enjoy!

1. Monday

It is 2:00 pm. Dr. Rufus and Tad welcome Willie to the control room and begin to explain why they have called on his computer expertise.

CARLENE: Tad, would you mind showing William to the control console? Do you go by 'Bill'?

WILLIE: 'Willie', actually. It looks like you're running a pretty powerful system. You must have two dozen new HP-UXs in here.

TAD: Twenty, actually, with about eighty Motorola VMEs for input–output control. Coffee?

WILLIE: Thanks, but it's a little late in the day for caffeine.

TAD [*pouring another cup*]: Suit yourself.

WILLIE [*looking around*]: Where does the funding for all this equipment come from?

CARLENE: The Department of Energy, primarily, but NASA, the NSF, and a few research universities are interested in what we're doing. We just might be on to something.

WILLIE: Something big?

TAD: Yes, well, no … something very, very tiny. We think we've found a new fundamental particle.

WILLIE: Wow, really? What is it?

CARLENE: We're trying to isolate the elusive psi-lepton. We've found something, but we're not sure whether we've found *it* or some unanticipated cousin. The particle we're observing was behaving exactly according to our theory, that is, until recently.

WILLIE: My particle physics is pretty rusty, but I think I remember reading something about the psi-lepton. It seemed like pretty speculative stuff, even for the fringes of particle physics.

CARLENE: Most of my associates have been quite skeptical, which is why we're taking such care in trying to understand our particle's anomalous behavior.

WILLIE: What's anomalous about it?

CARLENE: I'll let Tad fill you in. I have some results to look over and some more calculations to do. Tad, would you mind telling Willie about our dilemma, and then accessing the accelerator program files so he can get to work?

Dr. Rufus returns to her office, leaving Willie with Tad at the terminal near the control console.

TAD: Okay, so here's the story. The process we're running involves the ultra-relativistic collision of a uranium isotope and a heavy helium ion. We've set up B fields—uh, magnetic fields—to filter out every particle predicted to escape the collision except, so we expect, the psi-lepton. We have software in place to analyze the data from the detection devices, which track position and energy. The results were exactly what we were hoping for; it seemed undeniable that we'd observed a psi-lepton. Unfortunately, we haven't been able to replicate our results since early last week. Since then, although we can create a psi-lepton—again, assuming that's what it is—it vanishes long before our theory predicts.

WILLIE: So, your particle shows a faster decay rate than it should?

TAD: It's not that simple; it doesn't appear to be decay at all. We observe a second particle that appears out of nowhere, apparently annihilating the psi-lepton.

WILLIE: Out of nowhere?

TAD [*pointing to the monitor*]: Here's the weird data. (See Figure 1.1)

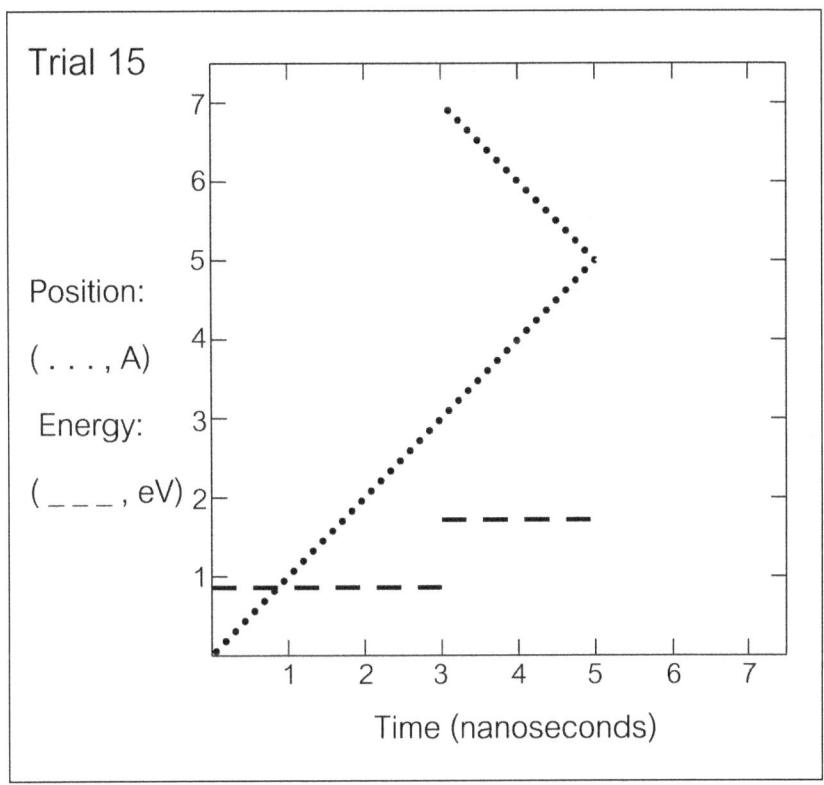

Fig. 1.1 The Unanticipated Results[1]

WILLIE: Tell me what I'm looking at exactly.

TAD: It's a standard position-versus-time graph that also indicates energy levels. The dotted line-segment emerging from the origin marks the position of the psi-lepton within the chamber at different times. The dotted line that emerges near the top of the graph at t=3 marks the position of the unexpected second particle. The horizontal dashed line marks the total energy within the chamber at the various times. The increase in energy at t=3 results from the presence of the second particle. Everything goes just as

---

1 To see an animation of any of the Monday illustrations online visit www.openbookpublishers.com/isbn/9781783740376#resources or scan the QR code.

expected from t=0 until t=3, but at t=3 we find the second particle in the chamber that appears to collide with the psi-lepton at t=5, after which there's no sign of either one. Trial 15 is our most recent trial.

WILLIE: But you weren't getting this result before the middle of last week, right?

TAD: That's right. Before then, we'd run eleven trials, and for each of those trials the psi-lepton was all by itself in the chamber and decayed at t=7, just as our theory predicts. Then we got four straight trials of who-knows-what.

WILLIE: And that's why I'm here? To tell you whether there's a bug in the accelerator's program that's giving you these strange results?

TAD: You've got it. Here are the accelerator program files. By the way, are you really a philosopher?

WILLIE: Well, I have a PhD in philosophy. I loved doing metaphysics, epistemology, and the like, but a full time job was hard to come by.

TAD: I guess you can't really expect DOE funding for that kind of stuff. I'll let you get to work.

Willie begins to study the programs that run the accelerator. After a few hours, he finishes and approaches Dr. Rufus in her office.

WILLIE [*knocking*]: May I?

CARLENE: Please, come in. Did you find anything?

WILLIE: I did, actually; there was a glitch. Since I was able to bypass the problem using an alternate code, my guess is that some kind of hardware problem cropped up last week. The new code is functionally equivalent, but the simulations run much more smoothly. As far as I can tell, the accelerator should work fine.

CARLENE: Excellent, Willie. Now we can see whether we really have the psi-lepton. Won't you stay for our first trial with your code?

WILLIE: Thanks, I will.

Dr. Rufus and Willie leave the office and walk back to the control console.

CARLENE: Tad, Willie thinks we're ready to go. Would you mind readying the accelerator?

Tad sits down at the console, entering the necessary commands.

TAD: All set, Professor. Shall I start it?

CARLENE: Please. (See Figure 1.2)

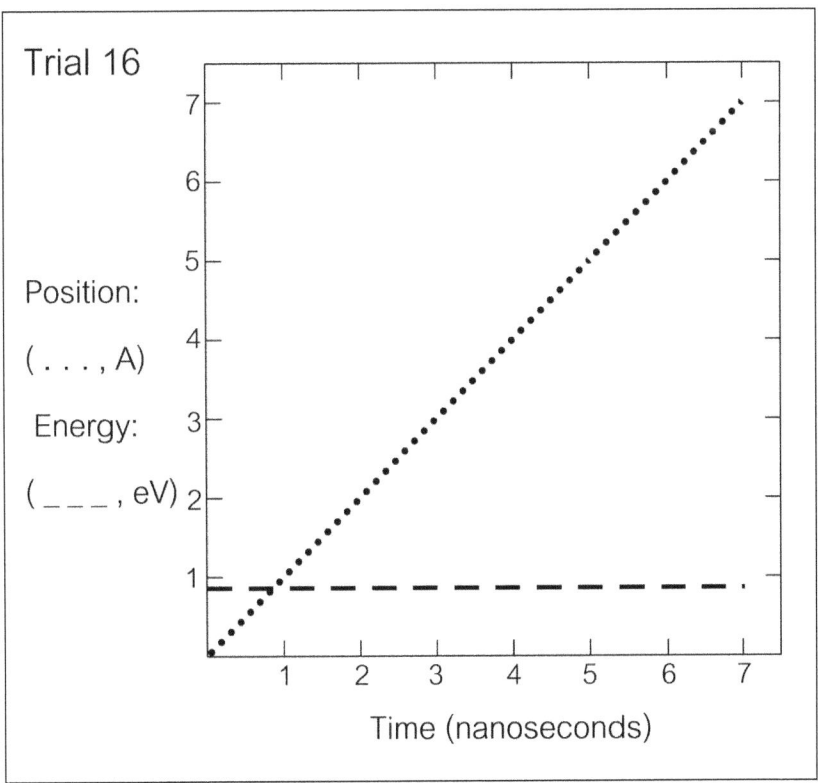

Fig. 1.2 The Anticipated Results

TAD: Willie, you've done it! Professor, the data indicate we had a perfectly stable psi-lepton in the chamber, just as our so-called speculative theory predicts. It's the same result we were getting before last week.

CARLENE [*containing excitement*]: Thank you, Tad, but don't be hasty. Right now I'm curious about what was happening before Willie bypassed that glitch. If the particle really is a psi-lepton, how can we account for the strange phenomena we were observing? Willie, do you know how the glitch was affecting the accelerator?

WILLIE: No, not really. Like I told you before, I suspect that there was some kind of hardware problem; something took place in the chamber when the program was run with the original code. Without pulling apart all of this beautiful equipment, that's about all I can tell you. If it'll help, though, I could probably figure out *when* the pesky event took place by restoring the original code.

CARLENE: It's not much, but every bit of information could be useful. Do you mind?

WILLIE: Not at all. Reinstating the original code should take only a second. Then I'll attach a diagnostic log that will show when the glitch kicks in.

Willie sits back down at the terminal and types for a few minutes.

CARLENE: Is that it?

WILLIE: That's it.

CARLENE: Tad, would you mind running the creation process again?

TAD: One psi-lepton, coming right up! (See Figure 1.3)

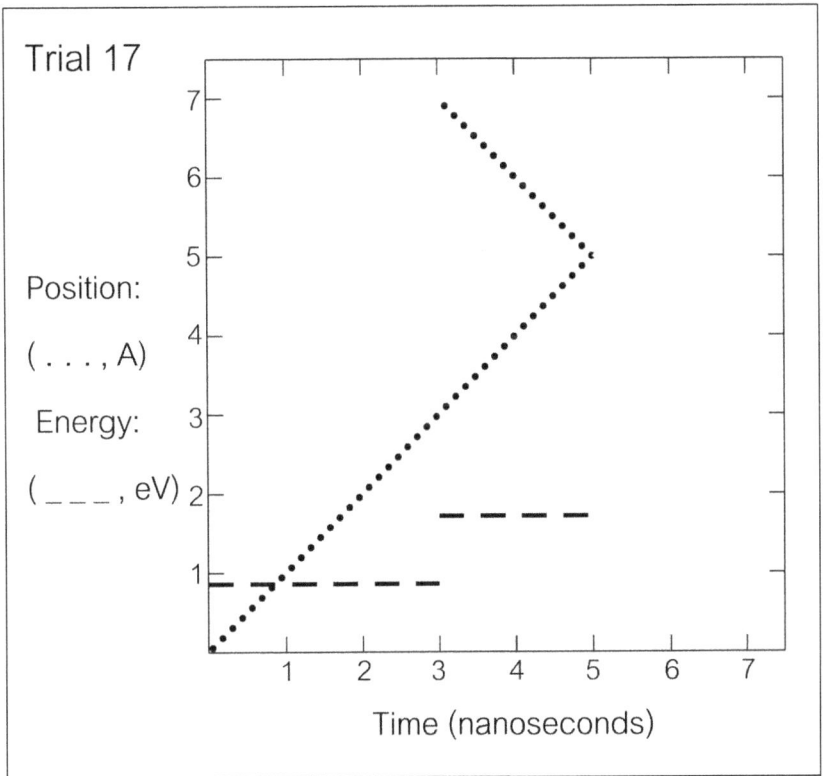

Fig. 1.3 Backwards Causation?

CARLENE: The anomalous particle is back. Once again, it originates at t=3 nanoseconds as it did in trials 13, 14, and 15.

WILLIE [*remaining fixated on his terminal*]: I know this is going to sound strange, but the event seems to take place at about t=5. Everything's normal until then.

TAD: That *is* strange. If Willie's right, the glitch didn't cause the second particle to appear. Its origin is still a mystery.

CARLENE [*tapping her finger on her nose*]: Hmm.

TAD: What are you thinking, Professor?

CARLENE: First, since we seem to be in control of this event, perhaps we shouldn't refer to it as a *glitch*. Second, in contrast to what you said, Tad, if the event doesn't take place, then the anomalous particle doesn't appear. When it takes place, however, the particle does appear. Maybe we should refer to the event as a *trigger*—it does, after all, now appear to be the cause of the second particle.

WILLIE: Backwards causation? You're suggesting that the trigger at t=5 causes the origin of the second particle at t=3?

TAD: What?! You're not serious, Professor.

CARLENE: It's the closest thing to an explanation for the appearance of the second particle we've come up with so far. Why couldn't it be a case of backwards causation?

TAD: Because that's wacky! The future can't cause the present or the past. Backwards causation is just the kind of confusing thing Hollywood takes up, not serious researchers.

Dr. Rufus raises her eyebrows in mock offense.

WILLIE: You know, Tad, what Hollywood seems to care about more than anything is entertaining and making money, thus the glut of paradoxical time-travel films. But serious researchers actually study this kind of thing; philosophers think about backwards causation and time travel quite a bit. In any case, backwards causation is the only real issue with our experiment. Carlene, if I've got you right, you believe that the trigger causes the anomalous particle to appear before the trigger is, well, triggered; then the new particle moves off and collides with the psi-lepton. No time travel, only backwards causation.

CARLENE: It would be imprudent to *believe* that or anything else at this point in our investigation, but I'm *considering* the idea.

TAD: Time travel or no time travel, it just doesn't make sense. The problems that come up in the time-travel movies aren't due to their time-travel aspect so much, but from the presence of backwards causation. Take *Back to the Future*, for instance: the time controls of the DeLorean are set in 1985, Marty ends up back in 1955, and young Biff chases Marty until Biff gets covered in manure as a result. The 1985 event caused the 1955 events. Okay so far, I guess, but say Marty time-travels to the past and permanently breaks up his parents' meeting, which he does only temporarily in the movie. If he had done that, he wouldn't exist any longer, but if he didn't exist, then he would never have been able to go back in time to break up his parents' meeting in the first place. It doesn't take long for things to get ridiculous.

WILLIE: I'll grant you that many time-travel films are hard to make sense of, but that's because the writers want to entertain. The logical absurdities slip in because the writers are more worried about dramatic effect and humor. But the problems in these films don't even begin to undermine the actual possibility of either time travel or backwards causation. Besides, there are other films that *are* coherent: *Terminator*, for one.

CARLENE: I'm afraid I'm not familiar with that film.

WILLIE: So, our society of the future is destroyed by cyborgs that were created by the company, Skynet. A devastating war between the machines and the humans takes place. During the war, the technology for a time-travel device is discovered. Events unfold, and one man and one terminator cyborg are sent backwards in time. The terminator is programmed to kill Sarah Connor, the mother of the leader of the human rebels, which will in effect prevent her from giving birth to her son.

CARLENE: That doesn't sound like a logical time-travel plot. How can the terminator succeed in killing the mother of a leader who later exists before she gives birth to him? Killing the mother would change the future; her son would never have been born and would never have fought the cyborgs, which the film evidently tells us he did. There appears to be a dilemma.

WILLIE: Well, you're right that there's a logical problem, but it's only with the *plan*. It seems the cyborgs didn't think it through. But what do you expect from cyborgs? The way things actually turn out, the terminator they sent wasn't successful; he didn't kill Sarah Connor. There's no suggestion that something both did *and* didn't happen. Sarah lives to give birth to her son, who later leads the humans against the cyborgs. It seems to me that as long as there's no hint that at one and the same time something both did and didn't happen, then the plot could be consistent. *Back to the Future* is a fun film, but it's hard to make sense of it if time is one-dimensional.

TAD: One-dimensional time? As opposed to what?

CARLENE [*interrupting*]: I'm going to have to cut this off. I want to get back to what's going on with the psi-lepton. Have a close look at this printout. Consider the possibility that at t=3 nanoseconds the second particle decays rather than emerges. Maybe t=3 nanoseconds is its instant of termination rather than its instant of creation.

TAD: Professor, just think about what you're suggesting. The particle can't decay then because it keeps going! It exists *after* t=3, until t=5. How can you say it's decaying at t=3?

CARLENE: Hold on, Tad. You shouldn't be so quick to ignore your own thoughts. Didn't you bring up time travel just a few minutes ago?

TAD: You can thank Willie for that.

WILLIE: Carlene, are you suggesting that the psi-lepton is time-traveling?

CARLENE: I think it's a candidate explanation for what we're observing. Wouldn't you agree, Tad?

TAD: A time-traveling psi-lepton? I really don't know whether that's worth our time.

CARLENE: Time travel could explain a lot of the data we've collected. Willie said that the trigger occurs at nearly the exact moment the two particles supposedly collide. What if there was only one particle? The trigger might cause the psi-lepton to reverse its temporal direction rather than directly and over a gap of time cause the birth of the second particle at t=3 nanoseconds. At t=5 nanoseconds, perhaps the psi-lepton ceases traveling from present to future, and begins traveling from present to past. What we've been thinking of as two particles could be just one particle, a psi-lepton traveling forward in time and the *same* psi-lepton traveling backward in time. Its lifespan would end when it decays at 3 nanoseconds, 7 angstroms from the origin.

WILLIE: In what way is that a better explanation than the hypothesis that merely posits backwards causation?

CARLENE [*holding a new printout*]: The graph only shows the data for the position and energy levels of the psi-lepton, but you can see in these supplementary data tables that in this trial the mass–energy, momentum, charge, spin—you name it—all have the predicted values. The data also indicate that there was no disturbance in the chamber's magnetic field anywhere near the supposed collision. If there had been a collision, then there should be some recorded disturbance in the chamber's magnetic field at t=5 nanoseconds.

WILLIE: I see; if the trigger caused the particle to turn around in time, to the effect that there was no collision, then we would have an explanation for the lack of magnetic field disturbance *and* for the origin of the second trace. We also don't need to posit any mysterious action at a distance between the trigger's occurrence at t=5 and a second particle at t=3. Wow, we might have just witnessed a case of actual backwards time travel!

TAD: Come off it, Willie! That doesn't mean anything. *Terminator* notwithstanding, time travel is a fantasy. It would generate all these crazy, impossible situations. I could go back in time and shoot my grandfather, but then if I had shot my grandfather, I wouldn't exist because my grandfather would never have fathered my father, and my father would never have fathered me. But then you'd have my grandfather lying dead in the street back in 1930, shot dead by a killer who never existed. Impossible!

CARLENE: Tad, how do you propose to explain the fact that the trigger seems to cause the anomalous particle to appear as well as there being no magnetic field disturbance after the supposed collision?

TAD: I don't know yet. Perhaps it's just a wild coincidence. Besides, you know those B fields can be pretty dicey sometimes.

CARLENE: Not just one wild coincidence, Tad, lots of them. We've run this experiment four times in the last few days and every time—excluding the time we ran it with the alternate code today—the so-called second particle has appeared.

WILLIE: Yeah, if we were always content to dismiss anomalies as coincidences, then I don't see how anyone could make progress in science. At least the time-travel hypothesis appears to explain away the coincidences.

TAD: But this is an extraordinary hypothesis, one that has its home in science fiction more than it does in actual science, and you're asking us to consider it on the taxpayer's dime to boot. Maybe a big dose of levelheadedness is in order.

CARLENE [*sighing*]: Look, guys, this talk really won't get us as far as more experiments will. We need to run more tests tomorrow. Do you mind returning tomorrow, Willie? We may need you to remove and restore the original code a few more times.

WILLIE: Sure, I don't have any pressing projects, and I think the Department of Energy can afford me for another day.

2. Tuesday

It is 9:00 am. Tad and Willie are in the lab talking over cups of coffee when Dr. Rufus arrives. It is obvious that none of the three slept well.

CARLENE: Good morning, gentlemen. Would one of you mind pouring me a cup? I was up all night thinking of ways to test our time-travel hypothesis.

TAD [*pouring Dr. Rufus a cup*]: That makes three of us, but maybe you had better luck than we did. I, for one, couldn't get over the idea that we're taking time travel seriously.

CARLENE: Please try to get used to it and think about ways to test our hypothesis. Where's the printout from yesterday?

Tad hands Dr. Rufus the cup of coffee and the printout of Trial 17.

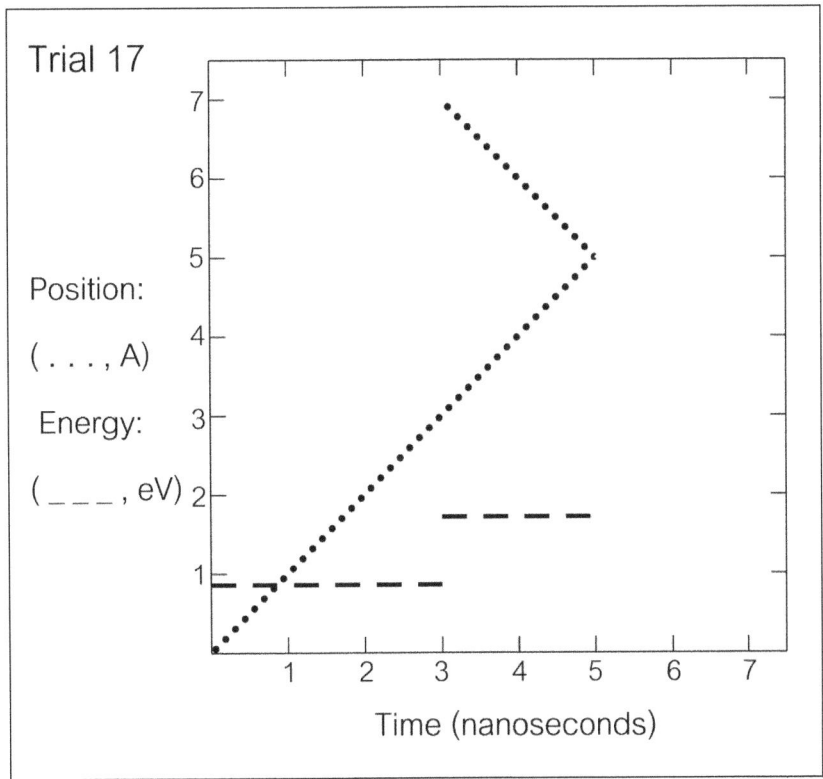

Fig. 2.1 A Time-Traveling Psi-Lepton?[1]

---

1 To see an animation of any of the Tuesday illustrations online visit www.openbookpublishers.com/isbn/9781783740376#resources or scan the QR code.

http://dx.doi.org/10.11647/OBP.0043.02

TAD: Let me get clear on what the hypothesis is exactly. You're claiming that—despite appearances—there's only one particle whose trajectory is plotted here, and that time-travel departure occurs at t=5. We're considering this hypothesis because it explains quite a lot; in particular, if true, it would explain both the lack of B-field disturbances at t=5 and the origin of what we thought was an anomalous second particle.

CARLENE: That's a fair enough account.

TAD: Okay, but I have a question about the supposed explanation of the second particle's origin and decay. Sure, according to the hypothesis, its *origin* is no longer a mystery; we created it when we knocked the uranium into the helium; it's a psi-lepton. But we still don't have an explanation of what's going on with that particle at t=3. That's what was really bothering us, what was so unexpected and left unexplained by the theory. So, like you're suggesting, what if that isn't the origin of the particle at t=3? According to the time-travel hypothesis, it's the time-reversed decay of the particle. Sure, I'll buy that, too. But I don't see that we have an explanation for why it decays.

CARLENE: Excellent, Tad. That's quite perceptive.

TAD: Yesterday we thought the time-travel hypothesis would account for the origin of the anomalous particle. What we're saying is that it doesn't exactly originate, at least not at t=3. That's neat, but if we're taking this idea seriously, there's still something that needs to be explained. In your version, Professor, the anomalous particle isn't coming into existence at t=3; it's decaying. But why should it do that?

CARLENE: It didn't occur to me until late last night, but I think that the hypothesis predicts the decay, too. Take a look at the duration of the lifespan of the particle in Trial 17. It originates at t=0, behaves normally until t=5 nanoseconds, and then travels backwards in time for two additional nanoseconds; its lifespan until it decays—how long it exists from its perspective—is seven nanoseconds.

TAD: And?

CARLENE: And of course you know that a duration of seven nanoseconds is precisely what our theory predicts for the lifetime of a psi-lepton. It's also how long our normal psi-lepton lasted in Trial 16 and some of our earliest trials.

TAD: Huh, that's clever, but I still don't know.

CARLENE: None of us knows much of anything at this point, I'm afraid. We're doing little more than speculating. What we need is more tests.

WILLIE: Maybe it would help to think about how the time travel we're considering differs from the backwards causation that you first proposed, Carlene; there's a big difference. Yes, the trigger seems to have some kind of backwards causal connection with the existence of the anomalous particle at t=3. It would be amazing if we could confirm even that much. But if the anomalous particle is the psi-lepton time-traveling, then the so-called two particles that the detector is recording are really one. That's crucial to the time-travel hypothesis; it's what takes the mysterious origin of the anomalous particle out of play. It also has some more fantastic-sounding consequences; for example, it implies that the psi-lepton is at two different places at the same time.

TAD: You're the philosopher; you figure it out. It doesn't seem like such a big deal that there's really only one particle; their sameness is easy to swallow. All I know is that there's no intrinsic difference, but they're elementary particles, so we should expect that. And yeah, they're in different places at the same time. So what? It's the problem of something in the future causing things in the past that bothers me. Whether they're the same particle seems trivial by comparison.

CARLENE: Let's drop the metaphysics for a moment and concentrate on what we really know. The path at the bottom part of the printout of our second trial from yesterday, Trial 17, is the theoretically predicted path of the psi-lepton. At all the relevant points in the chamber, it behaves exactly as predicted by my calculations, until the unexplained event occurs.

TAD: You mean the collision?

CARLENE: I'm not ready to decide whether it is a collision or whether it is a reversal in time, at least not yet. The point here is that, according to the data, we have a perfectly stable psi-lepton right up until the time of the trigger. The path at the top is where the confusion is coming from; all the data indicate it's a psi-lepton.

WILLIE: Even a perfect match wouldn't establish the time-travel hypothesis, but I'm not sure what would.

CARLENE: That's enough brainstorming for me; we're not getting any closer to trying a new experiment. What we need are some ways to manipulate our experimental set-up that might give us some more useful data.

Willie gets up and pours himself another cup of coffee.

WILLIE: Any ideas, Tad?

TAD: Well, like I said before, what was really keeping me up last night was the hypothesis that this trigger thing is causing the anomalous particle's presence. More than the very idea of the particle time-traveling, it seems weird that the cause of the particle's appearance happens later than the appearance itself.

WILLIE: Right, that's the big problem. You can bet that if we could find experimental evidence of backwards causation, it would send a shockwave through the philosophical world.

TAD: I'm sure. The big question is how to find the evidence.

WILLIE: Well, maybe there's some way to bilk the experiment; maybe we can stop the trigger after the second particle appears.

TAD: That might be tough given there are only a couple of nanoseconds between the particle's appearance and when the trigger kicks in.

WILLIE: But it might be feasible. It would take some tricky programming to optimize the processor's resources, and even then it would be a close call, but I might be able to.

TAD: What would it show us if you could?

WILLIE: It seems that even if we had time to stop the trigger, we wouldn't be able to since the effect of the trigger exists before the trigger itself happens, at least according to the time-travel hypothesis. If we do stop the trigger, we can rule out time travel and backwards causation as possible explanations for the second particle.

TAD: And if we don't stop the trigger, at least we'll have a lot to talk about. You'd probably argue that we would even have some experimental evidence that backwards causation is happening.

CARLENE: I think that would be a great way to proceed. If we can stop the trigger from occurring after the second particle appears, we'll have to consider that a strike against the time-travel hypothesis.

WILLIE: I'll get started on the program. We'll see how it goes, but I should be able to have it ready for trials today.

CARLENE: Excellent. Let me know when you're ready. I'll be in my office.

Dr. Rufus leaves Tad and Willie in the lab.

TAD: It seems to me that we don't even need today's experiment to disprove the possibility of backwards causation.

WILLIE [*typing at the console*]: How so?

TAD: Well, we already know we can change whether the second particle appears and whether the trigger occurs by inserting the new code. There's the practical matter of optimizing the use of the computer's resources to allow us to prevent the trigger in the time between the appearance of the second particle and the annihilation of both particles, but that's just a technological problem, and you're probably about to overcome it. Still, even if you don't succeed, we know it's *possible* to insert the new code some time after the second particle appears. So, assuming that backwards causation is at work, we can prevent the cause of the second particle's appearance after it's already appeared. But we shouldn't be able to do that, thus backwards causation isn't at work.

WILLIE: So, correct me if I'm wrong, but your argument goes something like this: If the trigger at t=5 causes the second particle's appearance at an earlier time, t=3, then it's possible to prevent the second particle's appearance at t=3 by switching in the new code at, say, t=4.

TAD: Right, that's what I said. We should be able to prevent expected effects by preventing expected causes. On Trial 16, the new code was in at t=4 and the second particle never appeared at t=3. It appears that the new code being in prevented the trigger, which apparently prevented the appearance of the particle.

WILLIE: But there's more: Setting the technological concerns aside, it's surely possible to switch in the new code at t=4 after the particle appears at t=3.

TAD: Correct.

WILLIE: So, if the trigger at t=5 causes the second particle's appearance at t=3, then it's possible to prevent the second particle's appearance at t=3 by using the new code at t=4 and it's also possible to switch in the new code at t=4.

TAD: Still correct, and there's the problem. In fact, it's impossible to prevent the particle from appearing at t=3 after it's already appeared at t=3. Preventing it would imply it didn't happen, but it would have already happened.

WILLIE: And, since it's impossible to prevent the appearance of the second particle after it's appeared, it's not the case that the trigger causes the second particle to appear.

TAD: Precisely. That's why it is logical to assume that backwards causation can't be the case and that the psi-lepton isn't time-traveling. So, what's the point of the experiment?

WILLIE: Well, the time-travel hypothesis might be false, but your argument doesn't show that.

TAD: Why not?

WILLIE: Your reasoning is invalid. I'll use another example to show you what I mean. Just think about this: It's possible for my coffee cup to be full right now, but it's also possible for it to be empty right now. Therefore it's possible for it to be both full *and* empty right now? That doesn't work.

TAD: Which is it?

WILLIE: What?

TAD: Your coffee cup.

WILLIE: Oh, truthfully? It's empty.

TAD: So much for your aversion to caffeine from yesterday. I'll get you some more.

Willie hands his cup to Tad, who goes across the lab to get some more coffee. Tad returns, handing Willie a full cup.

WILLIE: Thanks, Tad.

TAD: Okay, about your coffee example: I have to admit that it *is* impossible for the cup to be both full and not full right now; that really is illogical. But doesn't saying that it's possible that it's full and not full right now mean the same as saying that it's possible that it's full right now and possible that it's not full right now?

WILLIE: It may be tempting to interpret it that way; that's why your argument may have *seemed* to refute the hypothesis of backwards causation. As you've just acknowledged, however, that's a bad way to reason. It's obviously possible for the cup to be full right now and possible for it to be empty right now, but it's just as obvious that it's impossible for the cup to be both full and empty right now. More generally, I'm saying that, from possibly P and possibly Q, it doesn't logically follow that possibly both P and Q.

TAD: It just seems so much simpler regarding the coffee. How does all of this apply to the anomalous second particle?

WILLIE: It does so in just the same way. On the assumption that the trigger at t=5 caused the particle to appear at t=3, it was possible to prevent the particle's appearance at t=3 by inserting the new code, and it's also possible to insert the new code at t=4, but it doesn't follow that it's possible to prevent the particle's appearance once the particle appears. Once it appears it can't be prevented.

TAD: I see. So, you think we need to run the experiment?

WILLIE: Absolutely, we do. For starters, Dr. Rufus isn't going to pass up the chance for more data, especially not due to a philosophical argument—yours or mine. More importantly, we really need to see whether the anomalous particle ever exists without the trigger.

The three complete their tasks individually. After lunch, they reconvene to begin a trial with Willie's revision to the program.

CARLENE: Gentlemen, we appear to be ready for today's experiment. Willie, you're sure you have the accelerator program set to remove the trigger as soon as it detects the presence of a second particle?

WILLIE: Yeah, I was able to optimize the use of the processor so that the removal of the trigger can be accomplished in less time than the interval between the appearance of the second particle and the annihilation event.

CARLENE: Excellent. Shall we continue?

TAD: I'm starting the accelerator now. (See Figure 2.2)

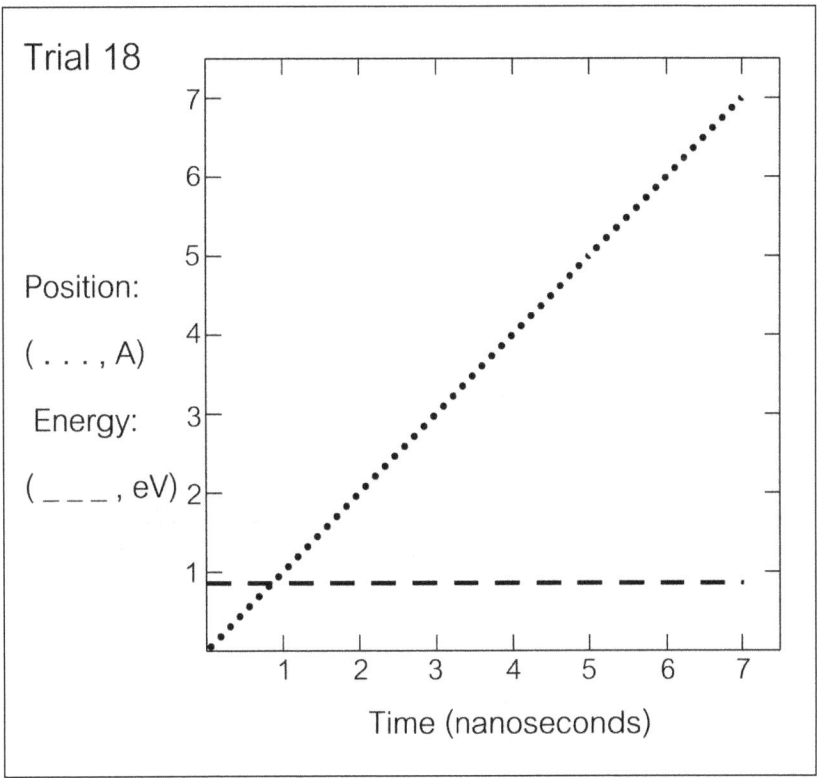

Fig. 2.2 Willie Slips Up

Dr. Rufus takes the printout of the latest trial and examines it. A wave of puzzlement passes over her face.

CARLENE: This is really puzzling. These data look something like the results when we removed the trigger altogether.

WILLIE: This is strange; the diagnostic log shows that the trigger was indeed removed upon detection of a particle in the chamber.

TAD: Isn't that just what it was supposed to do?

WILLIE: Yeah, so why don't we see evidence of the second particle?

CARLENE: Hang on, Willie; you said your program removed the trigger as soon as it detected *a* particle, right?

WILLIE: That's right.

CARLENE: At what time did that happen?

WILLIE [*checking the diagnostic log*]: The program removed the trigger at t=0.1 plus or minus 0.05 nanoseconds.

CARLENE: Hmm, just moments after the datum showing the creation of the psi-lepton.

WILLIE [*looking embarrassed*]: Um, I think I know what happened. When I rewrote the program, I set it to switch the trigger off when it detected *any* particle in the chamber, so when the psi-lepton showed up, the computer removed the trigger. We effectively repeated our Trial 16 from yesterday with no trigger at all. How stupid of me.

CARLENE: That would certainly explain these results, but it doesn't bring us much closer to establishing or falsifying our time-travel hypothesis.

TAD: Do you think you can modify the program so that it won't remove the trigger until *after* the anomalous particle appears?

WILLIE: Yes, I can do it, but it's going to take a little while; I don't want to blow it again. I'll need to make the program sensitive to the total energy in the chamber and set the threshold high enough to prevent the premature removal of the trigger.

CARLENE: Here's what we can do: one of our sponsors is hosting a conference tomorrow. Tad and I are planning to attend, so we won't perform any trials. There's also a conference luncheon. Willie, would you care to join us for that? It'll give us a chance to touch base and be ready for the following day. I can arrange a place for you.

WILLIE: That sounds good to me.

TAD: It's fine with me, too.

CARLENE: Excellent. It's been a long day already, so let's shut everything down. We won't be running any experiments until at least the day after tomorrow.

The three proceed to shut down the equipment in the lab. It does not take long for Tad to raise a question.

TAD: You know… I'm still having problems with the whole time-travel idea. Despite the evidence and the philosophy, I can't help feeling that this whole thing is foolish. It just isn't possible; you can't change the past; you can't time travel.

CARLENE: I agree that we can't change the past, but I fail to see how that makes time travel impossible.

TAD: It seems obvious to me. If you time-travel, you change the past; but since it's not possible to change the past, you can't time-travel. That seems like straightforward logic to me.

CARLENE: And how could a time traveler change the past?

TAD: Well, lots of ways. In fact, I don't see how one can possibly avoid changing history when traveling to the past. What if I really didn't like my grandfather? I wish I could have killed him, but he died before I had the chance, so I decide to go back in time to kill him before he even met my grandmother. Now, I only get one shot at going back in time; I want to make sure I can kill him, so I take all the latest weaponry with me.

WILLIE: Uh oh, this is sounding familiar. Keep going, Tad.

TAD: Since I have the best available arsenal, along with prior knowledge of my grandfather's whereabouts, it seems obvious that I can kill him.

CARLENE: Well, maybe you slip on a banana peel, or something else random like that happens? It would be a little like there being the misstep in Willie's program.

WILLIE: Thanks for the reminder.

CARLENE: Sorry, Willie.

TAD: Look, it doesn't even have to be anything as dramatic as killing my grandfather. I could step on a beetle or crush a blade of grass. Even my presence could slightly alter the flow of air. To me it seems impossible that I could time-travel to the past *without* altering the past. If nothing else, my tiny mass would slightly perturb aspects of the space-time continuum, resulting in some sort of change in the world. Since it's impossible to change the past, it seems that I can't possibly time-travel, and neither can our psi-lepton.

WILLIE: Tad, you're saying that you could kill your grandfather because you have what it takes to do so, but that you can't kill him because doing so would change the past. And you claim that this apparent contradiction rules out the possibility of time travel.

TAD: Yeah, absolutely.

WILLIE: Okay, but let's take another example. Let's say that I'm carrying a lot of groceries home, and I come up to the door, and I say, "Would you mind helping? I can't open the door". On the one hand, it seems that I said something true. On the other hand, it seems that I very well *could* have opened the door because I've done it hundreds of times.

TAD: How is that at all like my example?

CARLENE: I think I see how they're similar. Willie *can't* get through the closed door while carrying a full load of groceries, but he *can* get through the door when he isn't so burdened. He also could if he had longer arms, or if he had superpowers. In your example, Tad, we might have said it was possible for you to kill your grandfather if we hadn't taken into account the fact that he died of some other cause at a later time, that he fathered your father, and all that. But relative to all those things we *do* take into account about the case, no, you can't.

TAD: It seems to me that you're using some weird definition of 'can'. I just mean plain old 'can', as in 'possible'.

WILLIE: Dr. Rufus is right. When you say, "X can't happen", you normally don't mean that there's absolutely no logically possible way that X happens; there are even possibilities where the laws of physics are different. If 'can't' were to rule out all possibilities, we'd almost never use it. How 'X can't happen' actually seems to work is to indicate that there's no possible way that X occurs while other pertinent facts hold. What these facts are depends on the features of the conversation.

TAD: Okay, so what if 'can't' means there's no possibility of the event occurring while certain other facts hold? If I travel to the past, I still both can and can't kill my grandfather.

WILLIE: I don't think so. Either you can or you can't, but which one it is depends on what facts are taken as given. You can kill your grandfather given just the fact that you're well-armed, but not given the facts about how he actually died, say, of old age. You obviously can't kill him, given the fact that you didn't kill him. It doesn't really matter that you—his grandson—are the one trying to kill him; the same applies to anyone trying to kill your grandfather at a time he wasn't killed. Your scenario doesn't give rise to a contradiction. It would be a problem only if you both can and can't kill him given the same facts, but that's not the case.

TAD: But when I step out of my time machine and bend that blade of grass, I've changed the past!

CARLENE: I disagree. You caused the blade of grass to be bent, but that's not changing the past unless that blade of grass wasn't bent at that time in the past.

WILLIE: And it *was* bent if your story is going to be consistent. Don't try to tell us that it was as straight as an arrow at that time the first time, but was bent the second time around. That really is a contradictory story, but there's no reason to think that time travel is like that. If that blade of grass was bent at that time, then it was always bent at that time.

TAD: Well, if we can't change the past, what were those cyborgs thinking in *Terminator*—the movie you think is consistent—when they sent a terminator back in time to kill Sarah Connor?

WILLIE: That's one thing I didn't like about the film. There's no contradiction, but the cyborgs did reason poorly in thinking that they could prevent something they knew to have happened. But like I said yesterday, what do you expect from cyborgs? It's unfortunate that this provides the basic premise for the rest of the story. It's a rare time-travel film that acknowledges our inability to change the past, though there *is* the sci-fi gem, *12 Monkeys*.

TAD: What happens?

WILLIE: A man, James Cole, travels back in time. He's sent from 2035 to determine the origins of a deadly virus that plagued humanity in 1996, which drove the survivors underground. His goal is to take a sample of the pure virus to the future, to study and hopefully to discover a vaccine.

TAD: But if the vaccine is discovered, then the deaths of 1996 could be prevented after they occur, which leads to a contradiction.

WILLIE [*interrupting*]: Hold on, that's not what he was up to. The scientists knew they couldn't save the lives of those who were already dead; they wanted only to make a vaccine to prevent further deaths and to allow people to re-inhabit the surface of the Earth.

TAD: Okay, that makes enough sense, but it's starting to sound like I would *have* to step on that blade of grass as I step out of the time machine, which is too crazy even to consider.

WILLIE: Why is it so crazy?

TAD: Presumably there's nothing to grab my foot and push it down on that blade of grass; nothing would be forcing me to step on it. How can it be true that I would *have* to step on that blade of grass? If it had to be bent, it's not even clear how I could have caused it to be bent. It sounds like it would have to be bent no matter what I did.

WILLIE: The language is tricky here. When we say, "It would have to be bent" or "You would have to step on it", we're taking for granted that you *did* step on it. We're saying something pretty trivial, actually: given that you stepped on it, you have to step on it. We may as well say that, given the grass gets bent, the grass must get bent. These are really trivial claims.

TAD: But I wouldn't *have* to! No one would be forcing me!

WILLIE: Forget about the time travel for a second. Given that you will wash your coffee cup later, you have to wash it. There's nothing mysterious about that. It doesn't imply the existence of manipulative forces; it's just what you will do. And it's the same in the time-travel case: given that you stepped on that grass, you will have to step on it.

TAD: But I could decide to hop out of the time machine, rather than step down, and miss that blade of grass completely.

WILLIE: Well, maybe you *could*.

CARLENE: Wait a moment, Willie. Now it sounds like you *are* contradicting yourself. Which is it? Could he have hopped over the blade of grass or not?

WILLIE: Like I said, it's tricky. Don't forget about my trouble carrying the groceries. Could I have opened that door or not? It's a simple question to answer when it's clear what's being taken for granted. I surely said something true when I said, "I can't open the door", but there are other contexts where it's not taken for granted that I'm carrying a heavy, awkward load of groceries. Relative to these contexts, it's true to say, "Willie can open the door". After all, I would've needed nothing more than slightly longer arms, a tad—pardon the pun—more upper-body strength, or to be able to set down the groceries. Using 'could' in place of 'can' politely suggests that a change of the context is needed, that we should bring into play some more remote possibilities.

TAD: I'll just politely ignore your pun.

WILLIE: Look, Carlene, you asked whether Tad could have hopped out of the machine. Sure, he *could have*, but *can* he? This is a hard question to answer when it's not clear what's being taken for granted. If we suppose that Tad stepped on that grass however many years ago, that he *stepped down from* and didn't *hop out of* the time machine, then obviously he must step down. When he's faced with exiting the time machine, he can't hop out. If we don't suppose that Tad stepped onto the grass, however, if all we suppose is that Tad has the normal allotment of human capabilities, that there's nothing coercing him to do anything, and we also have no idea how things went, then he can exit the time machine in all kinds of ways. He may even decide the past looks scary and never leave the time machine.

TAD: Now it's starting to sound like there's *no* fact of the matter about what I'm able to do.

WILLIE: That isn't what I'm saying at all.

TAD [*interrupting*]: Hold it, Willie, don't say a thing. I'm too tired to keep this up. You haven't convinced me, but my head is starting to hurt. Professor, are we done here? If so, I'm heading home.

CARLENE: That sounds like a fine idea. I'll see you two at lunch tomorrow.

3. Wednesday

The three are seated at a table. The speaker, Dr. Twitchell, has just concluded a talk about the expanding universe. Lunch is being served.

CARLENE: It looks like we'll be gathering a lot more data about our mystery particle in the coming weeks. I might even say I *hope* it's time-traveling.

WILLIE: I'm still thinking about Dr. Twitchell's talk; it's always struck me as odd to describe the universe as expanding. What's it expanding into?

Dr. Rufus and Tad share skeptical glances.

TAD: I don't know about that, Willie, but something similar has been bothering me about our time-travel hypothesis. I'm wondering where the particle could go.

CARLENE: I don't understand.

TAD: If the psi-lepton reverses its direction in time, traveling from the present to the past, then we should acknowledge that the past exists, but that's not right. I mean, whatever happened yesterday, or even a moment ago, is done; it happened in another time, and that time no longer exists. This is now; the past—and the future, for that matter—don't exist. How can the psi-lepton time-travel if there's nowhere for it to go?

Dr. Rufus looks interested, but Willie is unimpressed.

WILLIE: That sounds like some arguments against time travel that I've heard from proponents of presentism.

TAD: And what's presentism?

WILLIE: Well, although it's defined somewhat differently by different philosophers, presentism is generally taken to hold that only what is present exists.

TAD: Okay, good, that seems pretty obvious to me. But I get the feeling there isn't general agreement that that's the way the world works.

WILLIE: Perceptive as always, Tad. Many philosophers believe in some form of eternalism, which contradicts presentism in holding that some non-present things exist. Eternalists believe that, in addition to what is present, both what is future and what is past exists.

TAD: So, you're saying that a presentist would claim that Albert Einstein doesn't exist, which is obvious, but that an eternalist would say he does exist?

WILLIE: Yeah, but the eternalist would concede that Einstein doesn't exist *now*.

TAD: That doesn't make any sense to me. How can you claim that something exists but doesn't exist *now*?

WILLIE: Well, I'm not claiming that, but the eternalist would just be saying that Einstein exists in the past but not in the present.

The server arrives with three lunches.

CARLENE: It actually seems obvious to me that Einstein exists, not now, but in 1905, for instance. We're talking about him, aren't we? Willie, I guess you would classify me as an eternalist. But I'm wondering, wouldn't the presentist admit that as well? Wouldn't he concede that, though Albert Einstein doesn't exist now, he does in 1905?

WILLIE: Well, strictly speaking, no. You have, however, come upon a key difference between presentism and eternalism: the eternalist might say that Einstein *exists* some years ago, but the presentist would say only that he *existed* some years ago.

CARLENE: That seems to be nothing more than word play.

WILLIE: In a way, it might be. I'm not entirely sure myself.

TAD: But what about my objection to our particle's being a time traveler?

WILLIE: There's a lot to be said. Let's make sure we have the argument right. You claim that if there's only the present, and if time travel requires a destination other than the present to travel to, then there's nowhere for a time traveler to go—better yet, no *when* for the time traveler to go.

TAD: Once again, Willie, you know exactly what I said.

WILLIE: Good. It seems to me that we have a few relevant options: we can give up our time-travel hypothesis while granting that the argument is sound, or we can give up presentism—there's still eternalism, after all—*or* we could hang on to both time travel and presentism and then try to uncover some flaw in the argument that doesn't demand that we reject presentism.

TAD: The argument seems foolproof to me, including the assumption that presentism is true. I lean towards giving up the time-travel hypothesis.

WILLIE: We know that's how you see it, Tad, but we might as well give it some more thought.

CARLENE: As tempted as I am to outright reject presentism, I suspect there's something wrong with Tad's argument. Sorry, Tad. It appears that if Tad's argument works—if it's sound, as you philosophers say, Willie—then all forms of time travel must be impossible. I doubt that presentism has such a strong consequence.

TAD: I think it does have such a consequence. Neither the future nor the past exists if only the present does. Any form of time travel requires the existence of a time other than the present, so if only the present exists, then any form of time travel is impossible.

CARLENE: Tad, think about the special theory of relativity, though; it tells us that at least a certain kind of forward time travel is possible. Time passes differently for material objects that take different paths through space and time; for example, a twin who makes a roundtrip at nearly the speed of light to a distant location ages much less than his twin who never leaves Earth. We actually know that this kind of time travel takes place; in one experiment, two atomic clocks—one placed on a plane and flown around for a few hours, and the other left on Earth—experience different amounts of time. The ability of atmospheric muons to reach the ground, the phenomenon of Thomas precession, and even the quantum-relativistic effects that give us the glitter of gold are all examples of well-understood physical phenomenon that involve this kind of time dilation. The important point, though, is that it seems that at least one form of time travel is very real.

WILLIE [*looking back at Tad*]: A hypothetical case makes the same point. If a woman were cryonically suspended for ten years before being revived, she would experience something like forward time travel. She might be convinced that it's 2010 when it's really 2020; she would be able to describe nothing about her trip but would have excellent recall of events immediately prior to suspension. That seems perfectly possible and like a case of time travel, so something must be wrong with your argument.

TAD: Seriously? That's not time travel!

WILLIE: Why not?

TAD: I'm not exactly sure. Even the atomic clock traveling at high speeds seems a little too mundane to be time travel, but I'll grant you that case, Professor. The deep freeze is a different matter.

CARLENE: I agree with Tad about the cryonic process. In terms of the physics, there's nothing interesting going on temporally.

The Northern wood frog, a species that regularly freezes solid during the winter, is not a species of time-traveling amphibian; there are intra-cellular and sub-molecular processes taking place even though little or nothing is happening at macroscopic levels.

WILLIE: Fine, but whether freezing and thawing is a way to time-travel is irrelevant. All I said, by the way, is that it was *like* a case of time travel. The point I want to make is that what's important about traveling is not that the destination be there when the traveler starts out, but that it be there when the traveler arrives. It doesn't matter whether the year 2020 exists when our woman is frozen, so long as it exists when she thaws. Similarly, it doesn't seem important that the arrival time as experienced by the clock on Earth exists when the plane starts to accelerate, but it's important that the arrival time exists when the plane returns. Tad's argument seems to overlook this particularly salient detail.

TAD: I don't follow.

WILLIE: Suppose you just graduated from high school. You and some friends have heard of an amazing theme park that's being built in Zimbabwe, so you decide to go there for an extended graduation trip, figuring that if you start walking when they hand you your diplomas, you'll get to the park just when it opens.

TAD: Okay, that's a pretty crazy story, but I'll play along. I guess my friends and I will need to do some swimming during the journey, too.

WILLIE: Good point. So, with your first step, and eventually your first backstroke across the Atlantic, you're traveling to the theme park, right?

TAD: Sure.

WILLIE: Well, the theme park doesn't exist yet, so you're traveling to a place that doesn't exist. If that's the case, then one of your premises is false, and your argument is unsound.

CARLENE [*smiling*]: Just a minute, Willie, I think you're trying to trick Tad. Perhaps the theme park in Zimbabwe doesn't exist yet, but the *space* where it will be exists, so Tad and his friends aren't really traveling to a place that doesn't exist.

WILLIE: Okay, okay, I'll concede that my example has that flaw, but—with Dr. Twitchell's talk in mind—we might suppose that our universe expands by creating new spatial locations; then you could take a trip to a place that doesn't exist. Of course, by the time you get there, it'll exist.

TAD: I still don't follow. If yesterday doesn't exist, then I can't travel there; if tomorrow doesn't exist, then I can't travel there. It would be like meeting Godzilla or traveling to the Fountain of Youth; it can't be done.

WILLIE: The difference is that the future *will* exist, and the past *did* exist. Godzilla and the Fountain of Youth never existed, don't exist now, and never will exist. You can be traveling to a spatial or temporal destination that doesn't exist yet; you just can't arrive until it does. When you're traveling to some time or place, you're engaging in traveling behavior, but you don't need to be simultaneously arriving anywhere.

CARLENE: Is that true, Willie? In the expanding-universe example, even if I want to travel to a region of space that doesn't exist yet, on the way there I would have to travel through—or arrive at— all the intervening space that does exist; the intervening space is what seems to make the trip possible. The analogy doesn't appear to hold up; if presentism is true, if the present is all there is, then there's no intervening time through which to travel.

WILLIE: But in a presentist's universe all of the intervening times will at some point be present.

CARLENE: That's not really the issue, is it? It seems that in order to travel to some place, we must travel through all the intervening locations; in space, these are readily available, but in a presentist universe there's only the present. Maybe we could, in manner of speaking, ride the present until a later time exists, but is that time travel? In the same way that the wood frog isn't time-traveling while it's frozen, the normal passage of time isn't time travel.

WILLIE: I would say that the normal passage of time *is* a form of time travel—albeit a limiting case of it—but you're right that we don't go out of our way to think of it like that. More has to be going on than just riding the present for an interesting case of time travel.

TAD [*interrupting*]: Professor, Willie admitted you were right!

CARLENE: I don't think Willie was finished, Tad.

WILLIE: I was going to point again to the twin-paradox case. The traveling twin occupies intervening positions in space and time on his way away from and on his way back to Earth, but what makes it time travel—genuinely interesting time travel—is how so much less time passes for him than for his stay-at-home twin and everyone else on Earth.

Tad quiets down. The server arrives, clearing the table and offering coffee. The offer is eagerly accepted.

TAD [*returning to business*]: I still think time travel is incompatible with presentism; actual time travel would require arriving at some non-present time, which presentists deny exists.

WILLIE: Look, how about this. A time traveler enters a time machine now and will arrive in 2020. The presentist should be fine with that.

TAD: There's still something screwy. This 'will arrive' thing bothers me. Until your time traveler arrives in 2020, it isn't true that she arrives at that destination. So, how can anyone *now* be time-traveling?

WILLIE: Sometimes our present-tense statements require for their truth what will or did occur to happen in a certain way. If our waiter is now placing arsenic in your coffee, isn't he *now* committing a murder, even though for that to be true it must also be true that, not knowing any better, you will drink the coffee and die? To be time-traveling now, you must be engaging in some sort of traveling behavior that causes that you did arrive or that you will arrive.

The coffee is delivered. Tad scrutinizes the contents of his cup.

TAD: Now it sounds like you're saying that being a time traveler doesn't really require that the past or future exist. It seems that all it requires is engaging in a certain behavior like traveling at high speeds or pulling the lever on a time machine. I have to admit that this sounds pretty reasonable. Traveling to ancient Greece implies that ancient Greece exists, but the presentist definitely denies that ancient Greece exists. Traveling at high speeds? Pulling a lever? These things seem perfectly consistent with presentism.

WILLIE: Then, maybe, we're in more agreement than I first thought. I was thinking of time-traveling to the past as consistent with presentism because I understood this as requiring only that the past *did* exist. You say that it also requires that the past exists and so see traveling to the past as inconsistent with presentism. Yet we can agree that I can be time-traveling even if presentism is true just by seeing time-traveling as a matter of engaging in the right kind of behavior. So, we seem to be in agreement on the important point that time-traveling is consistent with presentism.

TAD: So it seems.

CARLENE: Setting aside my worry from before, Willie, it might be helpful if you could tell us how some time-travel example fits with presentism.

WILLIE: Well, if one of you will outline one of your favorite time-travel plots, I'll show you how the same story can be told in a way that's consistent with presentism.

TAD: Okay, how about *Star Trek IV: The Voyage Home*. Kirk and the *Enterprise* crew travel back to twentieth century Earth, interact with some nobodies like us, find two humpback whales, and then take those whales back to the twenty-third century to stop an alien probe from destroying Earth.

WILLIE: Sure, I don't remember any obvious contradictions in that one. And remember, we said that presentists don't deny past- and future-tensed truths. So, suppose it's two days after the *Enterprise* crew picked up the whales sometime in the 1980s. At that time it would be true to say that there *was* a ship that picked up two whales two days ago, and in 300 some years that ship *will* appear near Earth.

CARLENE [*interrupting*]: I get the point. All you have to do is express the elements of the story in different tenses to make it fit with presentism. I'm still not convinced this isn't just a way of playing with the language.

TAD: Hang on, Professor, I just thought of something. Maybe the part of my objection about not having anywhere to go is dead, but I still think there's disagreement between presentism and the time-travel hypothesis. Willie, what about the causation our particle must be involved in if it's time-traveling?

WILLIE: I admit that there must be some strange causation; no matter which view—presentism or eternalism—is correct, backwards causation is troubling. But I don't think the type of causation we're worried about here raises any special issues for presentism.

CARLENE: Surely, though, if only the present exists, causation must occur only in the present.

WILLIE: Be careful to remember that causation isn't an event; it's a relation between events. Even though we sometimes talk about causation as if it's the kind of thing that happens in time, it doesn't actually. More importantly, there are countless apparent examples of causal relationships between present and non-present things; for example, the Big Bang caused the Earth, like every other material thing, to exist. Many of the things we do today will have effects in the future. If presentism is true, then it has to be able to explain how there could be any causal relationships at all, which is something presentism has to do whether or not time travel is possible.

TAD: Okay, so the apparent discrepancy between presentism and causation isn't limited to issues of time travel, meaning that the issue about causation doesn't really help us determine whether time travel and presentism are compatible.

The server returns and refills Willie's empty cup. Willie offers the server a thumbs up.

CARLENE: The thing that most worries me about presentism is scientific in nature, but we don't need to get into that right now.

WILLIE: No, go ahead, I'm curious.

CARLENE: Okay, so what worries me is that presentism seems to presume an absolute frame of reference that distinguishes what's real from what isn't; the present seems to consist of all and only those events simultaneous with right now. This is quite a departure from Einstein's relativity. The hallmark of relativity is that it doesn't include any notion of absolute simultaneity; simultaneity is a frame-dependent notion. If this is right, then presentism has to be false.

WILLIE: Yeah, that does seem to be a blow with respect to the truth of presentism, and—as I may have demonstrated earlier—I'm not exactly prepared to defend presentism until its dying day. This issue isn't really a problem distinctive of presentism, though; lots of our ordinary ways of thinking about time, space, and motion look to be at odds with relativity. Just as these ordinary ways of thinking have to be reconsidered in light of relativity, maybe elements of presentism need to be reconsidered too. Just as I'm not ready to say there's no such thing as motion and length, I'm not ready to conclude, well, that there's no time like the present.

The three share a chuckle.

TAD: Good one, Willie.

WILLIE: Just to be clear, even though I said "no time *like* the present", what I mean is that I'm not ready to conclude that times other than the present actually exist.

TAD [*interrupting*]: Lighten up, Willie. It was amusing; we get it.

CARLENE: Once again, this philosophical discussion is well and good, but if it *is* possible that the particle is time-traveling backwards in time, then it's up to science to show whether it's actually doing so. So, let's table the discussions for a while, until we have some more data to back up our ideas. With any luck—and some computer wizardry from Willie—we'll make some progress tomorrow.

4. Thursday

It is 7:00 am. Dr. Rufus and Willie are sitting at the computer console with cups of coffee, talking. Tad enters.

TAD: Wow, I thought that *I* was getting an early start!

CARLENE [*looking up*]: Oh, Tad, I didn't hear you come in; we're having an absolutely stimulating conversation.

TAD [*yawning*]: I think I need to have some stimulating caffeine before I can have any stimulating conversation.

WILLIE: Sorry, Tad.

TAD [*seeing empty coffee pot*]: Hey, don't you know that the first one in is supposed to make the coffee?

WILLIE: I do, and I did.

Tad looks from the empty coffee pot to Willie, who holds up his cup and tips it upside down to make the point.

WILLIE: It was good to the last drop.

TAD: Uh oh, what does that much caffeine do to a philosopher?

WILLIE: Hey, if the Department of Energy wants this work done, they'll have to lend me some of *their* energy. You can make the next pot; I think I saw some decaf in the cabinet.

TAD [*mocking horror*]: Decaf?!

CARLENE: Settle down, Tad. Willie's just teasing.

46 A Time Travel Dialogue

TAD [*recuperating*]: So, is that the program that turns off the trigger when two particles appear in the chamber?

CARLENE: No, it isn't. Willie's been trying to sell me on some new ideas for variations on the experiment.

WILLIE: After lunch yesterday, while you two attended the rest of the conference, I wrote the program that deactivates the trigger when two particles are present. I also wrote a program that is, in a way, the opposite; instead of turning the trigger off when the second psi-lepton appears, this program doesn't turn the trigger on until the second one appears. It's loaded right now.

TAD: Wait, tell me whether I have this straight. The trigger is initially off, but if two particles appear in the chamber, the program will turn the trigger on?

WILLIE: Now *you* are the master of summary.

TAD: Sure, Willie. But why are we bothering to run this program at all? I think it's obvious what will happen: regardless of whether the time-travel hypothesis is correct, if the trigger is turned off, then it'll look exactly like the trial without the trigger that we ran on Monday, Trial 16; the psi-lepton will live its life and decay normally.

CARLENE: Maybe, maybe not. We thought we knew what our data would look like the first time we isolated the psi-lepton, but we were certainly wrong then. In order to approach this problem scientifically, we need hard data, not speculation. Our knowledge of the psi-lepton's behavior is based almost entirely on our theory, and this trigger throws another unpredictable factor into the equation. I really don't think we can assume anything at this point.

TAD: What else could happen?

WILLIE [*pressing a few keys*]: Well, let's find out. (See Figure 4.1)

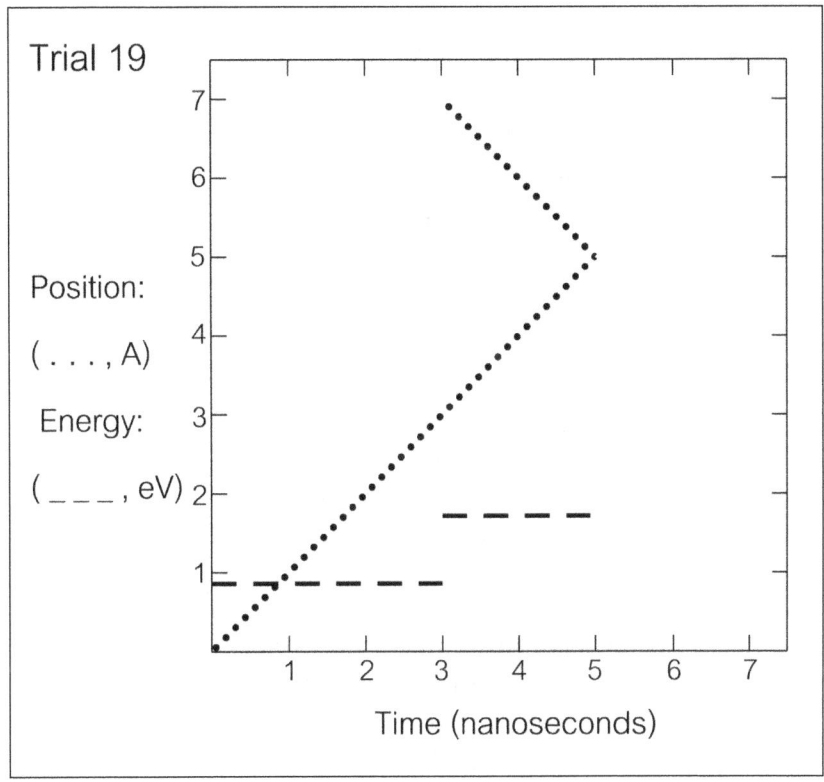

Fig. 4.1 A Causal Loop?[1]

CARLENE [*smiling*]: And *that's* why we should never assume that we know what the outcome of an experiment will be.

Tad grabs a printout of Trial 19.

TAD: How is this possible? These results look just like the other trials in which the trigger was turned on.

CARLENE: So, the so-called second particle, the time-traveling psi-lepton, turned on the trigger that caused the time travel.

---

1 To see an animation of any of the Thursday illustrations online visit www.openbookpublishers.com/isbn/9781783740376#resources or scan the QR code.

TAD: That shouldn't have happened even if the time-travel hypothesis is true. When the program was started, there was no trigger, and no trigger means no particle. Therefore, unless this just happened to be the one in eight bazillion times that another particle spontaneously appeared in the chamber, the experiment should have gone just like Trial 16.

WILLIE: Well, the trigger remaining off would have been consistent with our hypothesis; however, I believe that the actual results are consistent as well.

TAD: How can you say that? The second particle is clearly present in these results! According to your theory, the second particle shouldn't appear unless the trigger causes it, but the trigger was turned off in this program, so there was nothing to cause the second particle.

WILLIE: Not so fast, consider this: the anomalous second particle appeared, causing the program to activate the trigger; the trigger then caused the psi-lepton to change its temporal direction, thus traveling to the earlier time. The graph would look just like it does on the printout you're holding.

TAD: But there was nothing to cause the time travel.

WILLIE: Yes, there was; the trigger caused it.

TAD: But the trigger was turned off.

WILLIE: The trigger was *initially* turned off, but the presence of the time-reversed psi-lepton in the chamber turned it on.

TAD: But the particle shouldn't have been there!

WILLIE: But it was.

TAD [*containing frustration*]: But there was nothing to *cause* it to be there.

WILLIE: The trigger caused it.

CARLENE [*interrupting*]: Guys, you're going around in circles.

WILLIE: What a fortunate phrase! Yes, we're talking about the possibility of a causal loop here, wherein each event is among its *own* causes.

TAD: A causal loop? Oh please. What could have caused the loop itself? There's no reason that we should have a particle–trigger loop rather than no loop at all.

WILLIE: There seem to be a lot of possible causes; for example, by starting the accelerator, we could have caused the loop.

TAD: Wait, no, that can't be right. As we've already seen, lots of things could have happened after we started the accelerator; we got different results in the past. Starting the accelerator wasn't enough to guarantee that the loop occurred—assuming there's a loop at all—so how could it have been the cause?

CARLENE: Well, if we hadn't started the accelerator, then the causal loop wouldn't have occurred. Remember yesterday when we said that the Big Bang caused everything that's happened since? It's the same kind of thing; if the Big Bang hadn't occurred, none of the events since would have. But it's not clear that the Big Bang *guaranteed* any of the particular events that followed any more than turning on the accelerator guaranteed the causal loop.

TAD [*staring blankly*]: Okay, Professor, I see what you're saying, but I'm still not sure how some event could be the cause of another if it's possible for the first to occur without the second. Here's what I was getting at. Once the experiment started, there must have been some conditions that led to the occurrence of the loop rather than the ordinary life and decay of a single psi-lepton.

WILLIE: Our actual results were just one way that this could have gone; nothing about our set-up entailed that we would get the results that we did, with the trigger, rather than without the trigger.

TAD: Then how do we explain the results we got?

WILLIE: For starters, not all events are completely determined by initial conditions and the laws of nature alone; for example, quantum mechanics, on one standard interpretation, is an indeterministic theory. Given the laws of nature and the state of the universe at a time, there's only a certain probability that some possible state of the universe will follow, and that probability isn't 100 percent. If this is true—and if all events require an explanation—then we need some way of explaining undetermined events.

TAD: Which is?

CARLENE: We think about explanation in terms of causes all the time, Tad, often like this: some event causes another if and only if the effect-event is less likely to occur when the cause-event doesn't occur. During this trial we used Willie's new program, which starts with the trigger off but turns it on whenever two particles are detected in the chamber. If we had simply left the trigger off, then the causal loop probably—almost certainly—wouldn't have occurred, so it's reasonable to think that using this program—in addition to starting the accelerator, the Big Bang, and so on—caused the loop to occur. That seems like a pretty good explanation.

TAD: But we started with the trigger off; that's exactly how we've started several other trials. There was no difference in the initial conditions that led to the occurrence of the loop.

WILLIE: How can you say that the initial conditions were the same? Dr. Rufus *just* mentioned how the computer was running a different program and how that affected the possible outcomes.

TAD: The trigger was turned off initially, so I don't see how that makes any difference. In terms of probability, the chance of a second psi-lepton spontaneously appearing in the chamber is on the order of—well, the probability is essentially zero. But that's the only way another psi-lepton could have found its way into the chamber, and that's the only way the program could have turned the trigger on.

CARLENE: Willie, even though you make a plausible case, I have to admit to finding the causal-loop idea dubious.

WILLIE: Okay, maybe we should consider a different take on our results. You both seem to be assuming that every phenomenon in our universe admits of an explanation, is caused, and so on.

CARLENE: This is the best attitude to take in science.

WILLIE: Methodologically that may be correct, but theoretically that may be a little presumptuous.

Dr. Rufus raises her eyebrows in genuine offense.

CARLENE: Tread lightly.

WILLIE: Look, maybe it's a causal loop, which itself may be inexplicable. The Big Bang might be inexplicable, too, as well as why the universe is lawful in the manner that it is. Any event that quantum mechanics tells us was extremely improbable may also be inexplicable; citing low probabilities hardly explains why the improbable event occurred. Our causal loop—if that's what it is—is, I think, a good candidate for being an inexplicable sequence of events. Would it matter if it were inexplicable? Would you really suggest that we ignore the *results* in front of us because of a philosophical position on *explanation*?

TAD: Do you even know what 'tread lightly' means?

CARLENE: It's okay, Tad; Willie has a point. We do have a result that we need to contend with, and the best way is to get more results.

WILLIE: I think that part of the problem is that we're used to thinking about causes and effects in a linear fashion. Normally causes are independent variables, and effects are dependent variables; the causal chain doesn't usually link up to itself.

The three sit in silence for a bit.

CARLENE: So, in our supposed causal loop, every event is both its own cause and its own effect.

WILLIE: Yeah, I don't think any event in our loop really fits our usual assumptions about cause–effect relationships; I don't think we should try to identify the cause of our loop in terms of the ordinary linear relationships we're used to dealing with.

TAD: Maybe I'm too linear in my thinking. Then again, maybe these loops don't make any sense.

CARLENE: Right now I think we ought to try our first idea, the program that turns the trigger off when a second particle is detected. I believe this trial will really be the one to give clear evidence for or against the time-travel hypothesis. We have all the time in the world for data analysis and philosophical speculation.

WILLIE: I'm uploading the first program right now.

TAD: I still don't think causal loops make any sense. They make all sorts of ridiculous situations possible.

WILLIE: Such as?

TAD: Well, I once saw this movie called *Somewhere in Time*, but I don't remember much about it.

CARLENE: What do you remember?

TAD: I remember that there's a young man visited by an old lady in the 1970s, and she gives him a watch before leaving and saying only, "Come back to me". Many years later, the young man sees an old picture in a hotel of a beautiful actress that fascinates him. He does some research and finds a picture of that actress as an old lady, and he realizes that it's the same one who'd given him the watch. After talking to some weirdo who'd written a book about how to time-travel through self-hypnosis, the young man goes back in time to 1920. Once there, he finds the young actress, and they fall in love. Before returning to the 1970s, the young man gives the actress the watch that she will give to him when she's older. And so the history of the watch forms a complete loop, which makes no sense.

WILLIE: That's a pretty decent recall, I'd say. But what's wrong with the watch? It sounds to me like it had a completely consistent causal history: the man gave it to the actress in 1920; she carried it with her until the 1970s at which point she gave it to him; then he returned it to 1920 and gave it back to her.

TAD: What's wrong?! Nobody ever *made* the watch!

WILLIE: So?

TAD: So, that's impossible! Watches don't just appear out of thin air, Willie!

WILLIE: But it didn't; the watch's first appearance in 1920 was supposed to be the result of the time travel from the future.

TAD: And, as I've said all along, that's the underlying problem, the backwards causation. There's another problem, too. In order for the story to be consistent, the watch would have to be exactly the same when the actress first received it and when the man took it back in time to give it to her, right?

WILLIE: Definitely.

TAD: If that's true, though, then the watch wouldn't be able to age at all. The actress carried it around for fifty years or so; even if you assume the watch didn't rust or get scratched or whatever, its entropy would still have increased over that time.

WILLIE: Then the outside world must have expended energy to return the watch to its initial state. All that goes to show is that the longer the time between the arrival and the departure of the watch, and the larger it is, the more energy that's required to return it to the prior state. This could happen when the watch time-travels. Oh, the program is ready to go.

TAD: I still say there's no need to run this program.

WILLIE [*looking amused*]: Are you scared that you'll be proven wrong if we run this trial?

54 *A Time Travel Dialogue*

TAD: No, because there's no doubt in my mind that I'll be proven right; I'm just trying to save some money for the DOE. If we can turn off the trigger after the appearance of the second particle that was caused by the trigger—which we surely can—then we would obviously end up with a contradiction. That means that backwards causation is impossible, just as I've said all along.

CARLENE: Are you as confident about your predictions about this trial as you were about the last one?

WILLIE: Hey, we might *all* be surprised by the results. Tad, you ready?

TAD: Always.

The three gather intently. (See Figure 4.2)

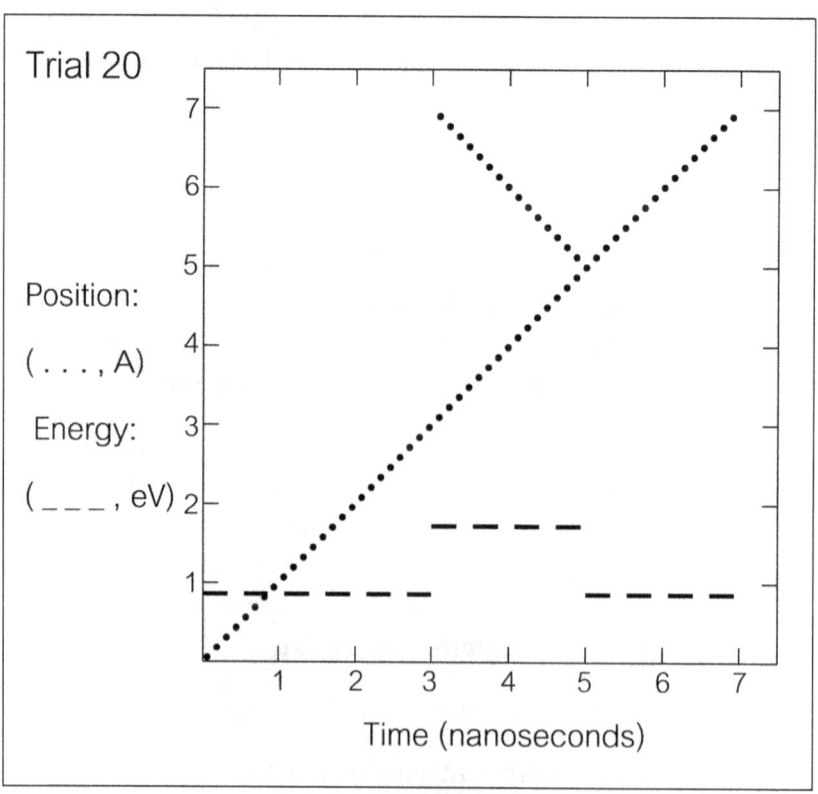

Fig. 4.2 The Really Weird Results

TAD [*studying the monitor*]: Well, I'm glad that we ran this trial, but as I expected it looks as though the time-travel hypothesis is shot. If the second particle really was the original particle traveling backwards in time as a result of the trigger, then it shouldn't have appeared. Now we have the anomalous particle appearing without the trigger. It lives its short life and decays just in time for the psi-lepton to rush past it. It looks like we're back to the drawing board.

CARLENE: Well, that's assuming the program ran as intended. Willie?

WILLIE: I'm almost positive that it worked. Based on the previous trial, the computer obviously can turn the trigger on or off in the interval of only a couple nanoseconds. Let me check. Yeah, the diagnostic log shows that the trigger was turned off after the appearance of the second particle.

> *Willie thinks to himself while Tad and Dr. Rufus quietly discuss the results.*

CARLENE: What do you think, Willie?

WILLIE: I have to admit that the possibility that the trigger caused the psi-lepton to travel backwards in time is now looking doubtful; there was no trigger-event, but there's still the path of what we thought was the time-traveling psi-lepton. Everything we've seen before suggests that this anomalous particle shouldn't have appeared, but it did, this time with a seemingly completely normal psi-lepton also in the chamber.

CARLENE: If the trigger wasn't causing the second particle, then why didn't it appear in Trial 18, or any of our earlier trials, when the trigger was never in place? Why now? For that matter, when there were the two particles, why did they both disappear? It never really looked like a collision or annihilation. But look here at the data tables: now we *are* seeing a slight disturbance in the magnetic field at both t=5 and t=7 nanoseconds, but there's no collision since the normal particle continues along its expected trajectory. We have the second particle, no collision, *and* the normal particle living its full life.

WILLIE: Do you think that time travel might still be involved?

CARLENE: Well, I have to agree with Tad that the time-travel hypothesis isn't looking so good at the moment. This was only one trial, though; there's more work to do. Who knows? Let's wrap this up for now. I'll see you fellows tomorrow.

5. Friday

It is 9:00 am. Tad enters the lab where Dr. Rufus and Willie are already halfway through their cups of coffee. Tad smiles smugly as he makes his way to the coffee pot.

CARLENE: What's got you so chipper this morning?

WILLIE: Do you mind, Tad? It's far too early for that.

TAD: Okay, so I've been doing a little research, and I spent last night brainstorming about our tricky little test subject, especially the result from yesterday.

CARLENE [*listening carefully*]: Please continue.

TAD: I still think that one-dimensional time travel is impossible. How can someone go back into the past and change it when that would change the future? So on and so forth. It would create direct contradictions.

WILLIE [*shaking his head*]: Well, no, Tad, because, even if someone were to travel to the past, the past would have already occurred the way it did; it would have included the arrival of the time traveler and his or her actions.

CARLENE: Do we really need to debate the grandfather paradox again?

TAD: Bear with me a minute. I finally found an answer that makes sense to me. Willie, do you remember on Monday when you mentioned the possibility that time isn't one-dimensional? The way around the paradox is to drop the idea that time has only one dimension!

WILLIE: Tad, I meant only to suggest that we can't even begin to make sense of films like *Back to the Future* without supposing that time is multi-dimensional. I meant to introduce to the discussion the *possibility* that time permits more than one timeline or branch of a timeline, but I never meant to suggest that time might *actually* be multi-dimensional. What are you trying to get at?

TAD: What if on one timeline my grandfather lives a normal life? He dies of old age, and then I decide to go back in time. My arrival is on a closely related timeline, but not the original timeline. And there you have it! I can kill my grandfather. The murder would have no effect on my birth because my birth isn't on this new timeline; my birth is only on the original timeline.

WILLIE: Yeah, I get the idea. It's a fun idea, Tad, really. But as far as the grandfather paradox is concerned—and as far as I'm concerned—your multiple timelines are gratuitous metaphysical musings.

TAD: Look, all week I've been searching the internet for information on time travel, and last night I came across this site, timetravelphilosophy.net, which a professor and his students assembled. It's got some really good stuff; it even puts forth some models of multi-dimensional time and discusses how the past could be changed. I can tell you where the particle is coming from!

CARLENE: You have my attention.

TAD: The website got me thinking: what if I decide to travel back to Willie's tenth birthday party?

WILLIE: That would be awesome. The time travel, I mean, not you at my party; my parents and I would have been more than a little creeped out to have this unfamiliar graduate student crashing my party.

TAD: Hey, it would be a blast; I even know how to juggle! Anyway, if I decided to travel back to your tenth birthday party, I wouldn't be traveling back to an event exactly like you remember, Willie; I'd be traveling to the party along a different dimension. The party I would visit would include my arrival. The original timeline—the one we're on right now—would branch as I arrived, with the original timeline going the way you remember it did, and another branch including my arrival. As one timeline plays out, I didn't arrive at the party; as the other plays out, I did. On this model, a time traveler to the past is bound to change the past.

CARLENE: Does this model really let the time traveler alter the past, though? It sounds like you would arrive at a very similar but distinct party, not the party of Willie's past.

TAD: I'm not sure that matters very much to our results.

WILLIE: Probably not, but I'm inclined to think that it could be the same party along the two branches. The party along each branch stands in causal and spatiotemporal relations to events that are identical along parts of the timeline prior to the branching, which seems to me to be sufficient for the identity of me with my younger self, my parents with their younger selves, and the continued existence of the party that started before Tad made his arrival. With one-dimensional time travel to the past, one person can be at two places at once, so why not one person along the original timeline and also along the alternate timeline? There could even be one event partially along the original line and partially along the branch; for example, this would be the case with the party.

CARLENE: I'm having a hard time visualizing this.

TAD [*taking out some paper*]: Okay, take a look. The t-axis represents normal time along the original timeline, and it indexes the normal times. A person born at (t_0, L_0) who never time-travels will live her life along the t-axis. Now, the L-axis keeps track of branches off the original timeline; L_1 will be the first branch. I'll graph the birthday example. Willie, what year were you born?

WILLIE: I was born in 1970.

60 A Time Travel Dialogue

Tad looks surprised. Dr. Rufus smiles briefly.

TAD: Hmm, you're older than you look. So, let's have t_0 represent 1970, and we'll tick off intervals of ten years. I was born in 1990, so my birth is here at t_2.

WILLIE [*interrupting*]: Tad, I didn't realize you were just a kid.

TAD: Come off it, Willie! Try to focus. So, if I decide in 2010 to depart in my time machine set to travel back 30 years, to your tenth birthday in 1980, my departure would be from (t_4, L_0), a point along the original branch; I would cross back and arrive at (t_1, L_1), which is 1980 on the arrival line. If I stay along that timeline for 40 years, I'll be about 60 at (t_5, L_1). Here, I've highlighted my life on the graph. (See Figure 5.1)

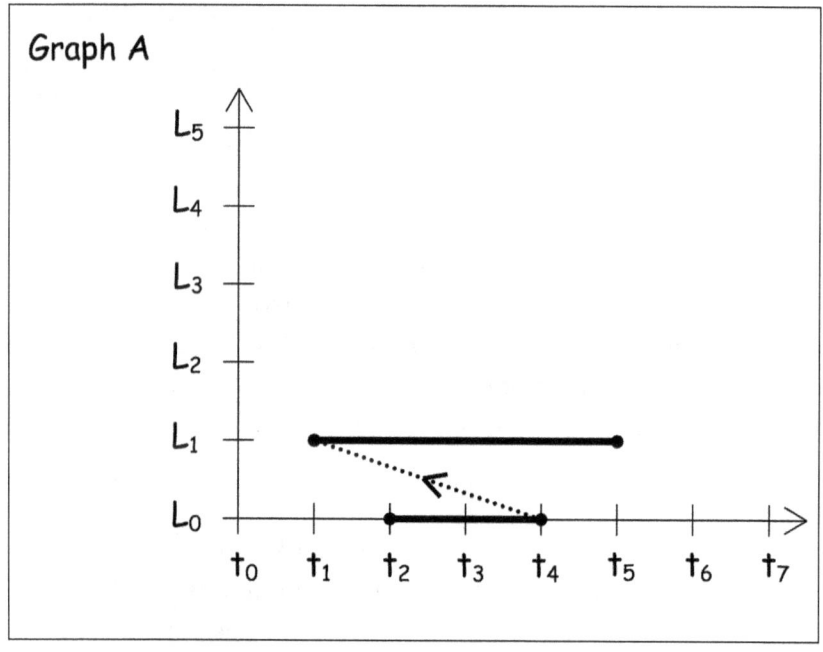

Fig. 5.1 The Life of Tad[1]

---

1 To see an animation of any of the Friday illustrations online visit www.openbookpublishers.com/isbn/9781783740376#resources or scan the QR code.

CARLENE: So (t_5, L_1) would correspond to 2020 along the arrival branch rather than the departure branch. You would lead your life, missing 2015 along the departure branch but being a part of it along the arrival branch.

TAD [*smiling*]: 'Departure branch', that's good. Don't you think this is really shaping up?

CARLENE: Why did you say you'd be about 60 in 2020 on the arrival branch, rather than exactly 60?

TAD: Well, you've got to consider whatever amount of time I experience during the trip, here along the diagonal.

CARLENE: Tell me a little more about this diagonal part of the trip.

TAD: Well, there are a couple of ways time travel could go: you could either *skip over* all the time between the departure and the arrival, or *move through* all the time between the departure and the arrival.

WILLIE: Right, that's the difference between *discontinuous* and *continuous* time travel. On a discontinuous trip, the time traveler leaves one time and instantaneously appears in another, without the traveler experiencing any time between the departure and arrival; this is the way things appear to go in *Back to the Future* when the DeLorean gets up to speed. On a continuous trip, however, the time traveler leaves one time and has to travel through the time and space between the departure and the arrival, to the effect that, as Tad suggested, time passes for the traveler between the departure and the arrival; the time machine in H.G. Wells' *The Time Machine* travels in this way. Time travel in *The Time Machine* is a one-dimensional, and we've clarified that time travel in *Back to the Future* only begins to make sense if it's multi-dimensional. So far, that's the Wellsian time machine's one-dimensional continuous travel and the DeLorean's multi-dimensional discontinuous travel, but what you seem to be describing, Tad, is multi-dimensional continuous time travel.

TAD: Right, like the boxes in *Primer*.

WILLIE: Oh, so you've seen it?

TAD: Yeah, the website on time travel suggested it; I watched it last night.

WILLIE: It's a good one. Tell Dr. Rufus about it.

CARLENE: Oh my.

TAD: *Primer* is a movie about two engineers, Aaron and Abe, who discover a way to travel backwards in time. They build boxes out of PVC pipe and who-knows-what; they turn these boxes on, and then later they can get inside and travel back to the time when the boxes were turned on. The catch is that they have to wait inside the boxes while they travel through all the time between getting in and getting out. As Willie said, the movie depicts multi-dimensional time, so they can change the past; they make big bucks buying stocks that they know will do well at the end of the day, and all kinds of other stuff.

CARLENE: So, figuratively speaking of course, you think that the psi-lepton is traveling in a *Primer* box?

TAD: You've got it, Professor.

WILLIE: This makes a certain amount of sense, but I'm not sure why they don't end up along L_2 or even $L_{1/2}$ for that matter.

TAD: I put the arrival branch along L_1 because I'm thinking of it as the branch on which the traveler stops traveling—and begins, once again, to experience the ordinary passage of time—after her *first* trip. If she left that branch, she would end up on L_2, and so on. When exactly along the arrival branch the traveler arrives depends on the settings of the time machine: 30 years back, 80 years back, or 80 years forward. Make sense?

CARLENE: Do all time travelers departing from L_0 arrive on L_1? What if they set their time machines to arrive at different times in the past?

TAD: Let's keep it simple: at most one time-travel departure for each line or branch and exactly one time-travel arrival for each branch. For every departure of a time traveler, there will be an arrival and this arrival is the beginning of the new branch off the departure line or branch.

CARLENE: What about the times along L_1 prior to your arrival at t_1? What events take place then?

TAD: As Willie hinted earlier, the events on L_1 before t_1 would be the same events that happened before t_1 on L_0; in general, the events before the arrival time on the arrival branch are the same events that happened before that time on the departure branch. On the graph, L_1 looks separated from L_0 both after *and* before the arrival time, but that's just a flaw of the visual representation. I intend for the events along L_0 prior to t_1 to be the causal and spatiotemporal antecedents of the t_1 and post-t_1 events along both L_0 and L_1.

CARLENE: Okay, Tad, you're right; this really is starting to shape up. But what does it mean for our research? Are you saying that every time the trigger happens, the psi-lepton travels to a different timeline? What observations might we expect with your hypothesis that time is multi-dimensional?

TAD: Well, that depends on which branch we're on. Are we on the branch where I show up at Willie's tenth birthday party or the one where I don't show up? Here's what I think: every trigger occurrence represents a departure of the psi-lepton. With multi-dimensional time travel, there are many more possible outcomes than with one-dimensional time travel. In multi-dimensional time travel, depending on which branch you're on, you're going to see completely different things on the printout.

WILLIE: It's like what I said my parents would observe at my tenth birthday party: along the departure branch there's nothing unusual, but along the arrival branch there's a creepy graduate student.

TAD: I'm glad you understand the model, Willie.

64   *A Time Travel Dialogue*

CARLENE: How about *our* observations? What's going on with the psi-lepton?

TAD: This is going to get tricky in a minute, but we can start off simply enough. If we're considering the psi-lepton as the time traveler on the departure line, let's make t_0 be zero nanoseconds and tick off intervals of one nanosecond; so t_5 is at five nanoseconds where the trigger occurs, if it occurs at all. Now, let's consider something like Trial 16, where the trigger doesn't occur. Here's what the graph would look like. (See Figure 5.2)

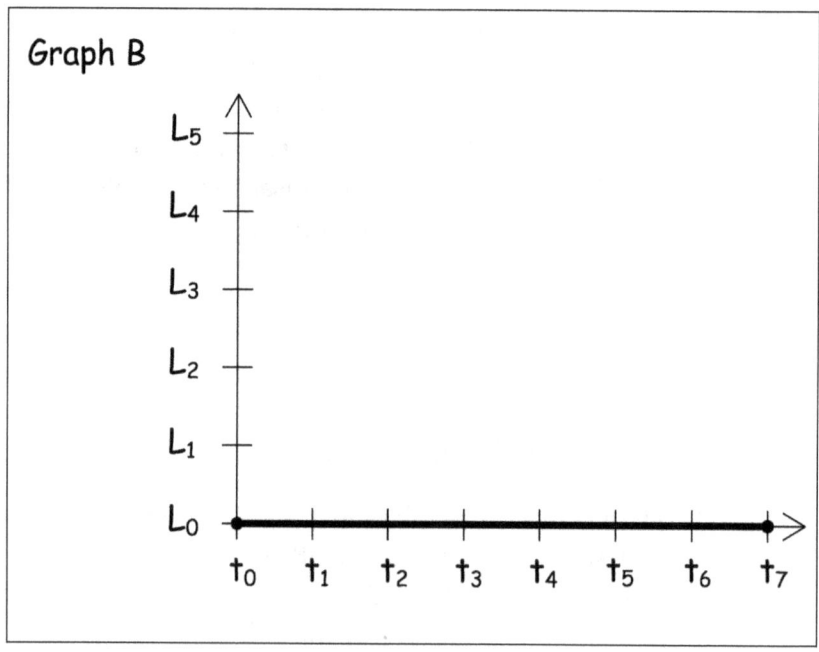

Fig. 5.2 An Ordinary Psi-Lepton: No Time Travel

WILLIE: No trigger, just normal psi-lepton behavior, living for seven nanoseconds and decaying at t_7 on the graph.

TAD: Right, but that's only the most straightforward case. Next we need to consider what the graph looks like when the trigger occurs at t_5. Keep in mind that we're making these observations from L_0. Also, remember we're working on the supposition that the trigger causes the psi-lepton to depart on a trip that's a case of inter-timeline time travel.

WILLIE [*interrupting*]: 88 miles per hour! It would look to us like the psi-lepton just vanishes, just like the DeLorean when it gets up to speed. The graph would show the psi-lepton existing only from t_0 to t_5. Why don't you draw that? (See Figure 5.3)

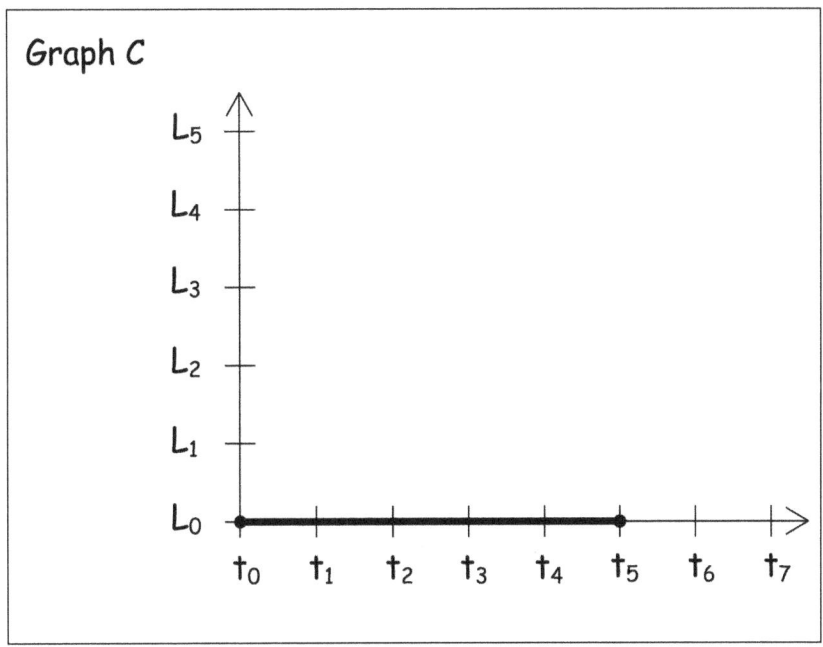

Fig. 5.3 A Time-Traveling Psi-Lepton as seen from the Departure Branch?

TAD: Calm down, Willie. And remember we're using the *Primer*-box metaphor, not the DeLorean. Think of it as Abe getting into the box. But you're right because, as far as we could tell, the particle's life would just end at five nanoseconds.

WILLIE: We haven't seen a trial with this behavior, have we?

TAD: Well, you haven't, but Dr. Rufus and I have.

CARLENE: Have we?

TAD: Yes, Trial 12.

WILLIE: Trial 12? You haven't mentioned that one; I assumed it was one of the early successes, a single particle that behaved according to your theory.

CARLENE: Of course, Trial 12! Willie, it was the beginning of our unexpected results. I had already dismissed it; we thought it was just a mishap of some kind since the results were never duplicated.

TAD: A mishap of some kind, but not exactly in the way we thought then. Here, Willie, take a look. (See Figure 5.4)

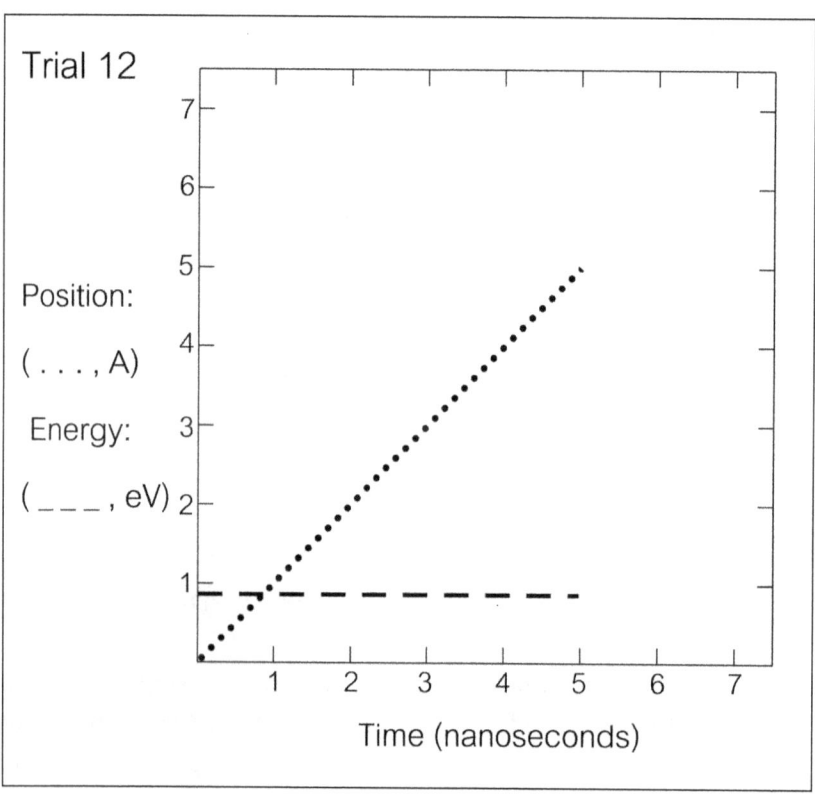

Fig. 5.4 The Disregarded Results

WILLIE: Are you sure the trigger occurred? It doesn't appear to have occurred because there's no second particle.

TAD [*pulling out the data tables*]: As you can see here, there was a B-field disturbance at t=5.

CARLENE: I remember now how strange I thought that was; it couldn't have been due to decay because the particle hadn't lived nearly long enough.

WILLIE: What does the disturbance have to do with anything?

TAD [*trying to contain excitement*]: Okay, let me explain what my theory says. First of all, the trigger is causing the psi-lepton to time-travel.

WILLIE: We've always had that as part of the hypothesis.

TAD: Give me a second, Willie. The occurrence of the trigger causes the psi-lepton to travel backwards across timelines. Remember that each step into the past changes the past. Traveling across timelines surely takes some energy, which has to come from somewhere, so the trigger draws it from the B field in the chamber; that's where the disturbance is coming from. During Trial 12, we observed a psi-lepton hit the trigger at t=5, at which point it got some energy from the B field, and it left our timeline. Always, the first question we need to consider is whether we're on the departure or the arrival branch. Here we're on the departure branch.

CARLENE: So, an equally important question is whether the trigger is *on* or *off* on our branch.

WILLIE: That's nice, Tad, but what about Trial 15? There are two particles, and there was no magnetic field disturbance. What about that?

TAD: There are a lot of different factors to consider. Depending on which branch we're on and—you're right, Professor—whether the trigger is on or off, there's a variety of possible outcomes. The behavior exhibited during Trial 15 is one such outcome. Let's take a look. (See Figure 5.5)

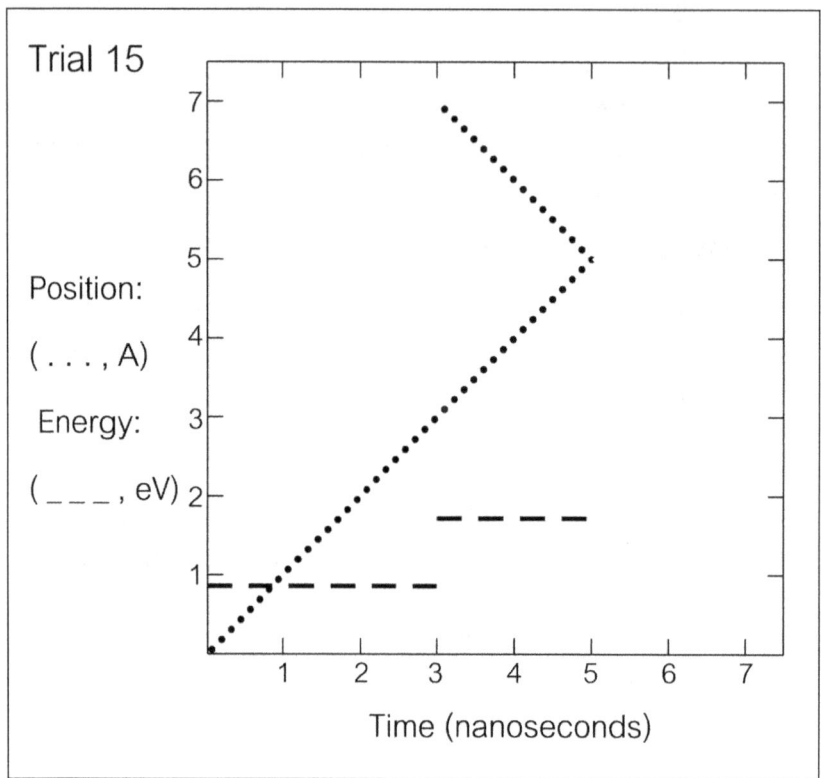

Fig. 5.5 A Time-Traveling Psi-Lepton as seen from the Arrival Branch?

WILLIE: I hope you've got something good.

TAD: Just hear me out. On our timeline the trigger is on; the psi-lepton travels normally until t=5, at which point it hits the trigger and leaves our timeline. Meanwhile, at t=3 there appears a psi-lepton from *another* timeline; on that timeline the trigger must have been on, too. Based on our previous time-travel hypothesis, the second path was the same psi-lepton traveling backwards in time; on my hypothesis, it's a psi-lepton from a *different* timeline, and it's traveling *forward* in time. It arrives at t=3 and moves towards the psi-lepton that originated on our timeline. Good so far?

WILLIE: So far, so good, I guess.

TAD: Now, you asked why there was no B-field disturbance during this trial. It's because the psi-lepton that arrived on our timeline decays! It lives the first five nanoseconds of its life on its timeline—its departure branch—and it lives the last two nanoseconds of its life on our timeline—its arrival branch—at which point it decays, here at t=5. The energy released from this decay supplies the energy for *our* psi-lepton to depart; it doesn't need to get it from the B field!

Dr. Rufus's eyes widen. Willie looks excited but nervous.

WILLIE: Remember when we were talking about your time-travel trip to my tenth birthday party? You said that after staying on the arrival branch for 40 years you would be *about* 60 years old. The psi-lepton has an extremely short lifespan. How is it living for five nanoseconds on its original timeline and two nanoseconds on our timeline when it must have spent some of its life *during* the diagonal trip across timelines?

TAD: I'm way ahead of you. We don't know what might be going on during an inter-timeline trip, but I think it's safe to assume that the physics is a nightmare. Now, the predicted lifetime of the psi-lepton is really a mean lifetime; it may live just a little bit less or a little bit *more* than seven nanoseconds. So, if the psi-lepton experiences less than two nanoseconds during its trip—and, for all we know, this is a possibility—then it's possible for it to live another two nanoseconds after it arrives.

CARLENE: This is clever, Tad, but I have one more question: if the psi-lepton that arrives on our timeline at t=3 nanoseconds is moving forward in time, why is it moving towards the lower part of the chamber, towards our psi-lepton? The path made sense when it was the same psi-lepton moving backwards in time—from its perspective it never changed direction in space—but I'm not sure about this.

TAD: That troubled me for a while last night, which is when I decided to take a break to watch *Primer*. I know this is kind of lame, but what if the psi-lepton has to turn around to get out of its *Primer* box to stop

70 A Time Travel Dialogue

time-traveling? What I mean is this: what if every time the particle stops time-traveling, its direction in space is reversed?

CARLENE: It's a stretch, Tad, but it makes enough sense to me, at least for now. What do you think, Willie?

WILLIE: I don't know. On one hand, it makes some sense of the data, like you said. On the other hand, why introduce so much tedium? Trial 12 aside, we had a pretty good and *simple* hypothesis about what we'd observed. Unless your theory can explain Trial 20 from yesterday, I don't see why we need it.

TAD: That's the thing, Willie; it *does* explain Trial 20! It explains *every* trial!

WILLIE: Okay, so show us. Here's the printout. (See Figure 5.6)

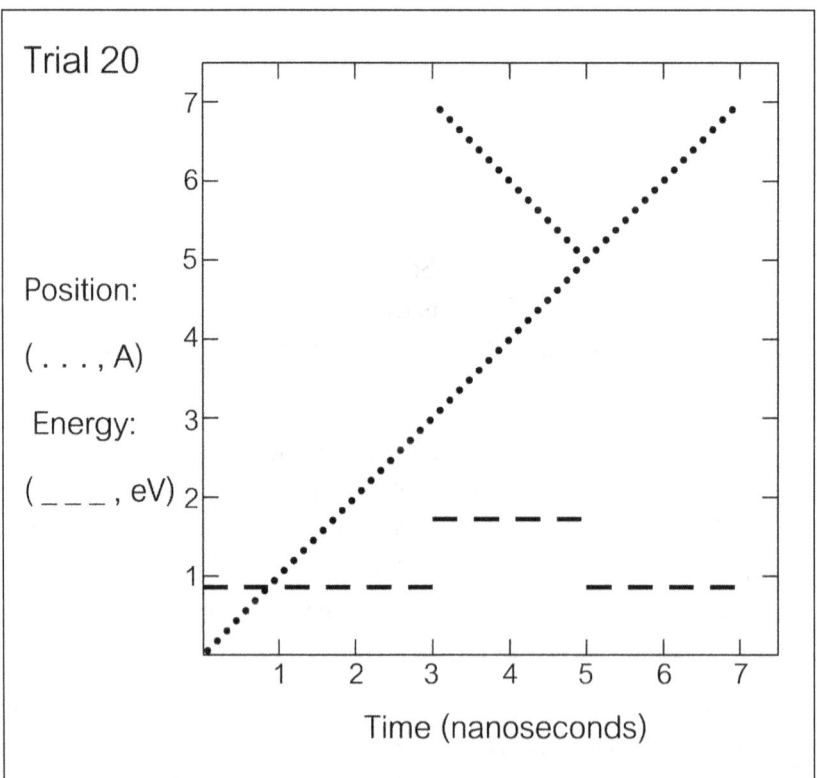

Fig. 5.6 Trigger-On Departure, Trigger-Off Arrival, as seen from the Arrival Branch?

TAD: After all I've just explained, it's a piece of cake. Remember that the program we were using during Trial 20 was set to turn the trigger off just in case a second psi-lepton was detected, and that the trigger was initially on. So, our psi-lepton is traveling along, and at t=3, the second psi-lepton appears from another timeline, which turns the trigger off. The second psi-lepton decays at t=5, and our psi-lepton—since there's no trigger anymore, remember—keeps moving until it decays as expected after about seven nanoseconds. The most important thing to notice here is that on the second psi-lepton's departure branch, the trigger never got turned off; otherwise it wouldn't have traveled to our timeline. The set-up on that departure branch would have had to be the same as ours—the same program running and all—but this is fine; all that means is that on the departure branch no second psi-lepton showed up to turn the trigger off.

CARLENE: What about the magnetic field disturbances we recorded yesterday?

TAD: That's easy. Since our timeline isn't a departure branch, when the psi-leptons decayed, they didn't have anything to feed their decay energy. The decay of the particle that arrived on our timeline at t=3 should have disturbed the B field at t=5, and it did. Our original psi-lepton—the one we created at t=0 on this timeline—should have decayed and disturbed the B field at t=7, and it did. Both disturbances are here.

Willie stares blankly for a moment before smiling.

WILLIE: So, on the departure *and* arrival branches of Trials 15, 17, and 19, the trigger was on, and we were on the arrival branch. But during Trials 16 and 18, the trigger was off, so there was no departure *or* arrival. During Trial 12, we were on the departure branch, which had the trigger on. And during Trial 20, we were on the arrival branch with the trigger off, but the departure branch had the trigger on. Damn, Tad, as crazy as it all seems, I'm impressed. I never really doubted the possibility of multi-dimensional time, but I never saw why it might be needed; it seemed ontologically irresponsible to assume it about our universe. I never thought there could be empirical evidence that spoke in favor of multi-dimensional time over one-dimensional time, but it looks like you found some. These models have to be taken more seriously than I ever thought. If we grant that the psi-lepton is time-traveling, our evidence favors it time-traveling multi-dimensionally rather than one-dimensionally.

TAD [*smiling*]: Thanks, Willie. But yeah, our recent trials are all evidence of time travel. The trigger always sends the psi-leptons back in time, and we're here sometimes to collect them and sometimes to watch them go.

WILLIE: I've got to take my hat off to you, Tad; you're now reasoning like Dr. Rufus was on Monday, looking for explanations that would make sense of the data. And I guess you have my hopes up a little that we're really witnessing time travel. I never liked multi-dimensional time; I never saw the point.

TAD: Is it too soon for me to buy a rifle and take after Gramps, or better yet Hitler? Too soon to check out the mid-cap stocks and make my fortune? Too soon to take a trip to the past in order to step around *any* blades of grass I well please?

CARLENE: Hold your horses, Tad. As incredible as all this seems, no one's time-traveling or even leaving this room until we get more results; we only have one example each of trials like 12 and 20. Are there any additional experiments we need to run?

WILLIE: Yeah, getting more results like Trials 12 and 20 would be great. But I want to look into something I just thought of. Tad, where's the printout for Trial 19? Ah, here it is. (See Figure 5.7)

TAD: What about it? The results are the same as Trial 15 and the others.

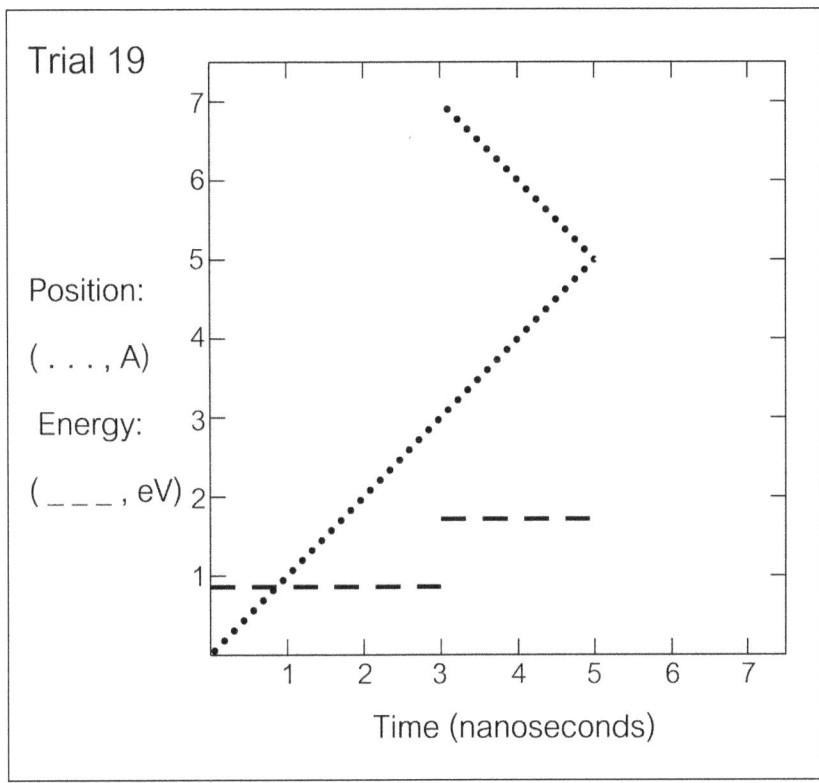

Fig. 5.7 A Challenge for Tad's Multi-Dimensional Hypothesis

WILLIE: The *results* were the same, but the experimental set-up was unique. We started Trial 19 with the trigger off, but the program was designed to turn the trigger on if it detected two particles in the chamber. Supposing time is one-dimensional, my explanation for the results was that this was one of two perfectly possible results, neither of which was determined by the conditions prior to three nanoseconds. The results were explained by a causal loop; the time-traveling psi-lepton in the upper part of the chamber was detected, which turned on the trigger, which sent the psi-lepton in the lower part of the chamber back to t=3 when it was detected.

TAD: Yeah, but that makes no sense to me. I don't believe in the possibility of causal loops just like I don't believe in the possibility of time-traveling to the past without changing it.

CARLENE: Oh, Tad, but that's the issue, isn't it? If there was no causal loop, what's your explanation of the results of Trial 19?

TAD: That's the beauty of my theory; I don't need a causal loop. The results of Trial 19 are consistent with the trigger occurring on the arrival and departure branches.

WILLIE: Be careful. The experimental set-up couldn't have been different on the departure and arrival branches; you said so yourself when you were talking about Trial 20. For Trial 19, the set-up was such that the trigger doesn't occur on the departure branch unless there are two particles in the chamber before three nanoseconds; the trigger was programmed to turn on if and *only if* the second particle was detected. Your model doesn't allow intra-timeline time travel, so how does the departure branch get the trigger turned on? It seems that a causal loop in one-dimensional time is the only way to make sense of the results. You realized yesterday that on your view the psi-lepton shouldn't have time-traveled during Trial 19. Unfortunately, it looks like it did.

TAD: Maybe we just can't explain these results. Maybe the second particle spontaneously appeared, and spontaneously decayed before it should have, and, hey, it's no worse than the causal loop, which you said might be inexplicable!

WILLIE: I suggested that the causal loop *itself* might be inexplicable, but the loop does allow for the results we got.

TAD [*grabbing at the air*]: Well, what if there was an infinite regression of branches, each one receiving a particle that turns its trigger on so that it can send its particle to the next branch

WILLIE: What about those branches, then? Are we on each one to infinity and beyond? One of the virtues of your model was that it allowed you to make predictions that could be observationally confirmed, but an infinite regression of branches? How do we confirm that? This is an *ad hoc* addendum to your hypothesis, Tad, and to be honest I felt similarly about your idea that the psi-lepton turns around in space after time-traveling. The rest was great, though; I'm sorry it doesn't work out.

TAD: I'm not sure what to say, Willie. I can't make sense of Trial 19, but you can't explain Trials 12 or 20.

WILLIE: We need more results. Carlene, what do you want to do next? Do you want the trigger to turn *on* only when there are two particles? I could also leave it to turn *off* only when there are two particles. We have a fair amount of choice.

CARLENE: Both Trials 19 and 20 raise interesting issues, and we've run those experiments only once each. Let's just leave the program from Trial 20 running for now. Maybe we'll get results like Trial 12; that would help out Tad's hypothesis.

TAD: I'd love anything in my favor. You ready, Willie?

WILLIE: Yeah, I'm all set. Don't get your hopes up for a repeat of Trial 12, though.

TAD: What do you think the chances are?

CARLENE: We're scientists. This is Jefferson National Laboratory.

TAD: Meaning?

CARLENE: We wait and see.

Notes

The Introduction makes reference to two nicely accessible introductions to the physics of time travel: J. Richard Gott's *Time Travel in Einstein's Universe* (New York: Houghton-Mifflin, 2001) and Paul Davies' *How to Build a Time Machine* (New York: Penguin, 2001). The philosophical work appealed to is David Hume's *Dialogues Concerning Natural Religion* (Indianapolis: Hackett, 1998).

Monday identifies issues raised in popular time-travel movies, touching briefly on the infamous grandfather paradox. The movies mentioned are *Back to the Future*, dir. Robert Zemeckis (Universal, 1985) and *The Terminator*, dir. James Cameron (Orion, 1984).

Tuesday builds on David Lewis's 'The Paradoxes of Time Travel', *American Philosophical Quarterly*, 13 (1976), 145-52, to address the grandfather paradox and other bilking arguments. Similar issues are addressed in Paul Horwich's, 'On Some Alleged Paradoxes of Time Travel', *Journal of Philosophy*, 72 (1975), 432-44; Kadri Vihvelin's, 'What Time Travelers Cannot Do', *Philosophical Studies*, 81 (1996), 315-30; and Jenann Ismeal's 'Closed Causal Loops and the Bilking Argument', *Synthese*, 136 (2003), 305-20. The movie discussed is *12 Monkeys*, dir. Terry Gilliam (Universal, 1995).

Wednesday focuses on presentism as the source of an objection to the possibility of time travel. The conversation includes many ideas from Simon Keller and Michael Nelson's 'Presentists Should Believe in Time-Travel', *Australasian Journal of Philosophy*, 79 (2000), 333-45, and also Phil Dowe's 'The Case for Time Travel', *Philosophy*, 75 (2005), 441-51. Ted Sider's *Four Dimensionalism*

(Oxford: Oxford University Press, 2002) includes an accessible presentation of the philosophical and scientific challenges to presentism; see pp. 11-52. Movie discussed: *Star Trek IV: The Voyage Home*, dir. Leonard Nimoy (Paramount, 1986).

Thursday delves into puzzling questions surrounding causal loops and their explanations. The example of a *jinni* (the watch) was drawn from the movie: *Somewhere in Time*, dir. Jeannot Szwarc (Universal, 1980). A discussion of these mysterious objects can be found in Gott's *Time-Travel in Einstein's Universe* (New York: Houghton Mifflin, 2001), pp. 20-24. Two short, delightful papers on this topic are Storrs McCall's 'An Insoluble Problem', *Analysis*, 70 (2010), 647-48, and Ulrich Meyer's 'Explaining Causal Loops', *Analysis*, 72 (2012), 259-64.

Friday raises the possibility that time is multi-dimensional, the idea that events take place along multiple timelines. For more fully developed models than the one sketched by Tad, see Peter van Inwagen's 'Changing the Past', *Oxford Studies in Metaphysics*, 5 (2010), 3-28; G.C. Goddu's 'Time Travel and Changing the Past: (Or How to Kill Yourself and Live to Tell the Tale)', *Ratio*, 16 (2003), 16-32; and Jack Meiland's 'A Two-Dimensional Passage Model of Time for Time Travel', *Philosophical Studies*, 26 (1974), 153-73. The book discussed is H.G. Wells' *The Time Machine* (New York: Tom Doherty Associates, 1992, first published in 1895). The movie discussed is *Primer*, dir. Shane Curruth (ThinkFilm, 2004).

Credits and Acknowledgements

A Time Travel Dialogue grew out of many years of teaching metaphysics to some exceptionally bright, eager, and hardworking students. From the outset, it has been a stimulating, collaborative project between myself and all my metaphysics students.

In the spring semester of 2000, I taught an undergraduate course in metaphysics at North Carolina State University that was structured around the question of whether time travel is possible. Traditional metaphysical issues, such as causation, identity over time and free will were addressed vis-à-vis this pedagogically unifying question. The class was broken up into groups which in turn wrote various parts of a dialogue whose structure was based on that of John Perry's *A Dialogue on Personal Identity and Immortality* (Indianapolis: Hackett, 1978). Four students from that class (authors Beth Ehrlich Slater, Kevin Harrison, Stuart Miller, and Nathan Sasser) and I continued to refine and rework this dialogue in an independent study course during the fall of 2000.

In the spring semesters of 2002 and 2003, I taught a total of four more versions of this course. The students from these courses were given the opportunity to add to the work that was already done by doing substantial rewriting of the existing dialogue or adding discussions of new topics. In the fall of 2002, three students from the spring 2002 classes (authors Steven Carpenter, Stephen Sutton, and Robert Todd) took the dialogue to the next level, incorporating graphs, the first discussion of presentism, and giving the dialogue a much more sciency feel. In the spring of 2003, author Laura Wingler contributed a substantial rewrite of the day on causal loops.

In the spring of 2006, students in my honors seminar on time travel were invited to add to the dialogue by incorporating discussion of multi-dimensional time. Author Diana Tysinger followed up in the spring of 2007 building on her work in the course by providing much of the original content on multi-dimensional time travel. This resulted in a slight reorientation of the entire dialogue. In the spring of 2011, author Kevin Martell did a first formulation of a detailed discussion of Friday's experimental results using Meiland's model of time. I have simplified the discussion by having Tad propose a simpler model that is very much in the spirit of Meiland's. Beginning in the summer of 2013, Gray Maddrey took the dialogue to publication by building more detail into the discussion of the Friday results; he also worked through the entire dialogue to make the prose more fun to read and in line with the new results and discussion he added to Friday.

Though many, many other students made important contributions besides these primary authors/writers, special mention is due Evan Johnson for his suggestions on the cover design, Chris Streshe for his work on presentism, David Schlorff for his work on the grandfather paradox, Jay Hodges for his work on causal loops, and Kristoff Kleiner for his work on multi-dimensional time. Jay and also Melissa Schumacher have gone on to do their own research on time travel and remain trusted advisors. Allyson Hutchinson was a terrific philosophical ally on all topics temporal, metaphysical, and web.

The philosophical community of time-travel enthusiasts is large and growing. There are too many to list all of those who have shaped my own thinking and, many a time, brightened my day, but Sara Bernstein and John Roberts lead the way. Larry Blanton and the University Honors Program at North Carolina State have been a constant source of support for my teaching. Thanks also to the North Carolina State Department of Philosophy and Religious Studies for its support of my teaching and financial support for this project as it moved to publication. Michael Pendlebury, the department head, and his administrative staff of Ken Peters and Ann Rives are terrific colleagues and friends. Thank you to my wife, Ann Carroll, for patiently working with me to make the current images. Three referees for Open Book Publishers provided very useful comments. Special thanks to Brad Skow for detailed and insightful comments on the entire manuscript. Open Book's vision of open access publishing drew me in, but the efficiency and expertise of their staff has been absolutely first rate; Open Book has been every bit as excited about *A Time Travel Dialogue* as my students and I are.

The dialogue is a work by my students. I have been closely involved by virtue of being the one to expose the students to the metaphysics of time travel and in doing the substantial editing needed to make all the different student contributions fit together into a coherent whole. Still, the students have done the bulk of the work. The authors/writers are a group of especially bright students, several of them former North Carolina State honors students, many now with graduate degrees and successful careers. It has been a joy to be back in touch with them all. The combination of their cleverness, their devotion, and their sense of humor has produced something remarkable.

John W. Carroll
July 3, 2014

This book need not end here...

At Open Book Publishers, we are changing the nature of the traditional academic book. The title you have just read will not be left on a library shelf, but will be accessed online by hundreds of readers each month across the globe. We make all our books free to read online so that students, researchers and members of the public who can't afford a printed edition can still have access to the same ideas as you.

Our digital publishing model also allows us to produce online supplementary material, including extra chapters, reviews, links and other digital resources. Find *A Time Travel Dialogue* on our website to access its online extras. Please check this page regularly for ongoing updates, and join the conversation by leaving your own comments:

> http://www.openbookpublishers.com/isbn/9781783740376

If you enjoyed this book, and feel that research like this should be available to all readers, regardless of their income, please think about donating to us. Our company is run entirely by academics, and our publishing decisions are based on intellectual merit and public value rather than on commercial viability. We do not operate for profit and all donations, as with all other revenue we generate, will be used to finance new Open Access publications.

For further information about what we do, how to donate to OBP, additional digital material related to our titles or to order our books, please visit our website.

www.ingramcontent.com/pod-product-compliance
Lightning Source LLC
Chambersburg PA
CBHW070558160426
43199CB00014B/2545